QUANTUM PHYSICS AND THEOLOGY

Quantum Physics and Theology

An Unexpected Kinship

JOHN POLKINGHORNE

Yale University Press New Haven and London

Scriptural quotations throughout this work are taken from the New Revised Standard Version (NRSV).

Designed by James J. Johnson and set in Janson Roman by Tseng Information Systems, Inc.
Printed in the United States of America.

Library of Congress Cataloging-in-Publication Data
Polkinghorne, J. C., 1930–
Quantum physics and theology : an unexpected kinship / John Polkinghorne.
p. cm.
Includes bibliographical references and index.
ISBN-13: 978-0-300-12115-5 (cloth : alk. paper)
1. Religion and science. 2. Quantum theory. 3. Theology. I. Title.
BL265.P4P58 2007
261.5′5—dc22 2006024231

A catalogue record for this book is available from the British Library.

The paper in this book meets the guidelines for permanence and durability of the Committee on Production Guidelines for Book Longevity of the Council on Library Resources.

10 9 8 7 6 5 4 3 2 1

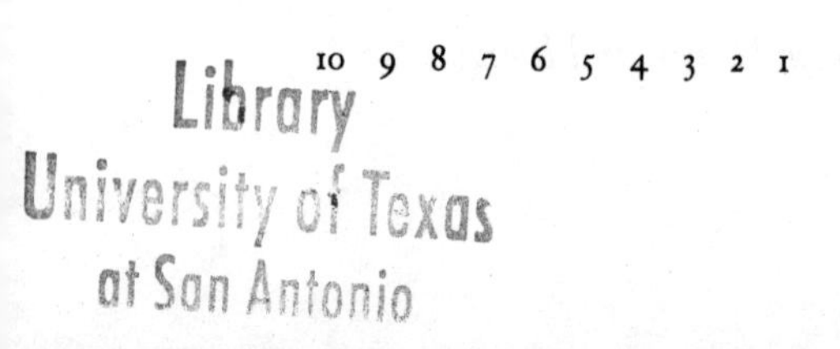

To Ruth

Can two walk together, except they be agreed?
—Amos 3:3 (AV)

Contents

Preface

I have always found it difficult to decide what to choose for the title of a book. In the case of this short volume, I was tempted at first to go for 'Quantum Theology'. However, I rejected that idea because the choice of such a title might have given the disagreeable appearance of trying to appeal to a style of discourse that one might call quantum hype—the invocation of the peculiar character of quantum thinking as if that were sufficient licence for lazy indulgence in playing with paradox in other disciplines. It is certainly true that quantum theory serves as an illustration of the fact that reality often turns out to be stranger than we could have thought. Its example warns us against the error of trying to make commonsense expectation the measure of all things. Yet the strangenesses encountered at different levels of reality have characters that are idiosyncratic to those levels, and no facile kind of direct transfer is possible between physics and theology.

Another possible title might have been 'Christian Science', but Mrs Mary Baker Eddy got there first and preempted that option. That was a pity, for it would have been a suitable rubric under which to present this book, which is an essay with a single controlling theme, namely that, contrary to an all-too-common misjudgement, it is not the case that theology and science are chalk and cheese, a matter of airy opinion compared with solid fact. Nor does the essential difference between them lie in a contrast between belief on the basis of submission to an unquestionable authority and belief based on grounds of rational motivation. Quite the contrary, for there are significant degrees of cousinly relationship between the ways in which science and theology conduct their truth-seeking enquiries into the nature of reality, though there are also certain obvious differences that arise from the contrasting characters of the dimensions of reality with which each is concerned. The thesis of an underlying truth-seeking connection between science and theology appeals strongly to someone like myself, who spent half a lifetime working as a theoretical physicist and then, feeling that I had done my little bit for science, was ordained to the Anglican priesthood and so began a serious, if necessarily amateur, engagement with theology. I do not discern a sharp rational discontinuity between these two halves of my adult life. Rather, I believe that both have been concerned with searching for truth through the pursuit of well-motivated beliefs, carefully evaluated.

The thesis that science and theology are friends and not foes, and that they share in the use of certain common ways of approaching the search for understanding, is one that is quite often discussed in the writings of those who might be categorised as scientist-theologians, persons whose intellectual formation has been similar to my own, in that while it started in professional science, the continued pursuit of understanding has eventually led them to theology.[1] This theme has certainly been a feature of my own writing, perhaps most clearly expressed in my Terry Lectures, in which I endeavoured, amongst other matters, to identify five points of comparison and similarity between the physicists' search for an understanding of the nature of light and the theologians' search for an understanding of the nature of Jesus Christ.[2] (The scheme of argument laid out on that earlier occasion is summarised in the course of chapter 1 of this book.)

In this slim volume I want to return to this theme in considerably more detail, laying out a whole series of comparisons between modern quantum physics and the centuries-old theological pursuit of an understanding of the nature of God, pursued in a Christian context. I think that the comparison of science and theology, and the defence of the thesis that both are best understood as leading to a critical realist account of what is the case, has to be conducted largely by the analysis of specific examples, rather than by attempted appeal to very general kinds of argument. The technique that I shall follow here will be to take a number of features of the way in which rational enquiry is conducted and its results evaluated, in each case illustrating the point being made first by an example drawn from physics, and then by an analogous example drawn from theology. In each of these paired discussions, I shall seek to describe what is involved in a concise fashion that makes the structure of thought clear without overburdening

1. For a survey of the work of three scientist-theologians (Ian Barbour, Arthur Peacocke and the present writer), see J. C. Polkinghorne, *Scientists as Theologians*, SPCK, 1996.

2. J. C. Polkinghorne, *Belief in God in an Age of Science*, Yale University Press, 1998, ch. 2.

the account with unnecessary detail. If the passages relating to physics were all taken together, they would give a kind of skeletal account of a great deal of what has been going on in fundamental physics in the last hundred years. However, a person for whom the story of modern physics was itself the main focus of interest would undoubtedly want to be given a more fleshed-out account, of the kind that one can find in a number of excellent guides to modern physics written for a lay readership. If the passages relating to theology were all taken together, they would give a skeletal account of significant aspects of the development of Christian thinking over many centuries, but a reader whose principal interest lay in that subject would doubtless want more detail, of the kind that introductory textbooks on historical theology provide for the student. My purpose in restricting myself in both cases to the barebones of what is involved is that such a treatment allows the homologies between the two rational structures to be discerned most clearly. Birds and human beings look very different on the outside, but when their bone structures are revealed and analysed one can see that wings and arms are morphologically related to each other. In an analogous fashion it is possible to exhibit the basic similarities of truth-seeking strategy that exist between physics and theology through a procedure of attending to accounts stripped down to their essentials.

It would be disingenuous not to confess that I also have a second purpose in mind in writing this book. I hope that its analysis may help some theologians to recognise more clearly the cousinly relation that their discipline bears to science and so, perhaps, to be encouraged to engage with the latter a little more seriously than many of them have seemed inclined to do. In addition, I hope that what is written may help some of my scientific colleagues to take what theology has to say with a greater degree of seriousness than many of them display. In the scientific community, the adjective 'theological' is sometimes used pejoratively to refer to a vague or ill-formulated belief. I believe this usage to be very far from the truth. It saddens me that some of my colleagues remain unaware of the truth-seeking intent and rational scrupulosity that characterise theological discourse at its best. When I read the writings of some of the high-profile scientific proclaimers of atheism, I find a degree of ignorance of the intellectual content of serious biblical study and theological reflection that is not altogether different from the scientific ignorance displayed by those who send out papers with titles such as 'Einstein was wrong'. If scientists are to reject religious belief, they should do so with their eyes open and after a proper consideration of the serious intellectual effort that has been exerted in theology over many centuries of careful enquiry. I do not pretend that so short a book as this can, in itself, be an adequate presentation of the riches of theological thought, but I hope that it may indicate something of the truthful intent with which theologians seek to speak of the infinite mystery of God, and that this will lead some enquiring scientists not to set aside too hastily insights that are worthy of their careful attention. Perhaps this proffered hors d'oeuvre might encourage some to sit down to a more substantial meal.

Acknowledgements

I thank the staff of SPCK and Yale University Press for their work in preparing this book for press. In particular, I thank Simon Kingston for helpful comments on an early draft of the manuscript.

This is the first book of mine whose proofs have not been subject to the eagle-eyed scrutiny of my late wife, Ruth. I gratefully dedicate this book to her memory.

John Polkinghorne
Queens' College
Cambridge

CHAPTER ONE

The Search for Truth

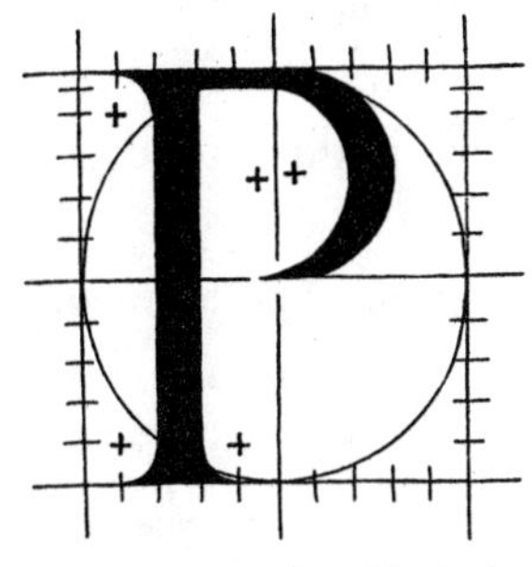

EOPLE sometimes think that it is odd, or even disingenuous, for a person to be both a physicist and a priest. It induces in them the same sort of quizzical surprise that would greet the claim to be a vegetarian butcher. Yet to someone like myself who is both a scientist and a Christian, it seems to be a natural and harmonious combination. The basic reason is simply that science and theology are both concerned with the search for truth. In consequence, they complement each other rather than contrast one another. Of course, the two disciplines focus on different dimensions of truth, but they share a common conviction that there is truth to be sought. Although in both kinds of enquiry this truth will never be grasped totally and exhaustively, it can be approximated to in an intellectually satisfying manner that deserves the adjective 'verisimilitudinous', even if it does not qualify to be described in an absolute sense as 'complete'.

Certain philosophical critiques notwithstanding, the pursuit of truthful knowledge is a widely accepted goal in the scientific community. Scientists believe that they can gain an understanding of the physical world that will prove to be reliable and persuasively insightful within the defined limits of a well-winnowed domain. The idea that nuclear matter is composed of quarks and gluons is unlikely to be the very last word in fundamental physics—maybe the speculations of the string theorists will prove to be correct, and the quarks, currently treated as basic constituents, will themselves turn out eventually to be manifestations of the properties of very much smaller loops vibrating in an extended multidimensional spacetime—but quark theory is surely a reliable picture of the behaviour of matter encountered on a certain scale of detailed structure, and it provides us with a verisimilitudinous account at that level.

Theologians entertain similar aspirations. While the infinite reality of God will always elude being totally confined within the finite limits of human reason, the theologians believe that the divine nature has been revealed to us in manners accessible to human understanding, so that these self-manifestations of deity provide a reliable guide to the Creator's relationship with creatures and to God's intentions for ultimate human fulfilment. For the Christian, this divine self-revelation centres on the history of Israel and the life, death and resurrection of Jesus Christ, foundational events that are the basis for continuing reflection and exploration within the Church, an activity that the community of the faithful believes to be undertaken under the guidance of the Holy Spirit. Revelation is not a matter of unchallengeable propositions mysteriously conveyed for the unquestioning acceptance of believers, but it is the record of unique and uniquely significant events of divine disclosure that form an indispensable part of the rational motivation for religious belief.

Both these sets of claims for truth-bearing enquiry are made in conscious conflict with much of the intellectual temper of our time. In many parts of the academy the movement broadly called postmodernism holds sway. It emphasises what it sees as the uncertain basis of human knowledge, a vulnerability to challenge that results from the inescapable particularity of perspective imposed by the need to interpret experience before it can become intelligible and interesting. The necessary cultural context of language is held to imply that there is no universal discourse, but only a babble of local dialects. The grand modernist programme of the Enlightenment, alleged to be based on access to clear and certain ideas that are unquestionably acknowledged to be universally valid, is asserted to have been no more than the imposition of the perspective of white male Western thinkers, treated as if such attitudes were a non-negotiable rule for all. According to postmodernism, casting off these modernist shackles liberates twenty-first-century thinkers into being able to accept a creative plurality of ideas, thereby enabling participation in a conversation in which nobody's opinion has a preferred priority.

We certainly need to acknowledge that rational discourse is a more subtle matter than Enlightenment thinkers were able to recognise. Yet in the eyes of its practitioners, science does not at all look like a free market in ideas of an eclectic kind. We shall attempt to specify its character in more detail shortly, but one must begin by considering how it actually progresses. The more extreme postmodernists would challenge the use of that last verb, but can one really suppose that the concepts of the helical structure of DNA and the quark structure of matter do not represent clear advances in understanding, intellectual gains that have become a persisting part of our understanding of the world? These ideas evolved under the irresistible nudge of nature, and not as fanciful notions whimsically adopted by the invisible colleges of molecular biologists and particle physicists respectively. Once the famous X-ray photographs taken by Rosalind Franklin had been seen and understood, there could be no doubt that DNA was a double helix. Once the data on hadronic structure (the patterns found in the properties of particles that make up nuclear matter), and the results of deep inelastic scattering (a particularly penetrating experimental probe), had been collected and assessed, there could be no doubt that fractionally charged constituents lay within protons and neutrons. Of course, interpretation was necessary—raw data such as marks on photographic plates are too dumb to speak of structure directly—but the naturalness of the interpretation, and its confirmation through a continuing ability to yield more understanding in the course of further lengthy investigations, is sufficient to convince scientists of the verisimilitudinous character of their theories. It is difficult for those not involved in scientific research to appreciate how difficult it is to discover theories that yield persistently fruitful and elegantly economic understanding of extensive swathes of experimental data, and therefore how persuasive such understandings are when they are attained. For those of us who were privileged to be members of the particle physics community during its twenty-five-year struggle to understand nuclear matter—an activity that eventually led to the Standard Model of quark theory[1]—the enterprise had precisely this convincing character. The experimentally driven investigation, often proceeding in directions quite different from the prior expectations of the theorists, was no indulgence in the construction of pleasing patterns, but it was the hard-won recognition of an order in nature that is actually there.

A just account of science lies, in fact, somewhere between the two extremes of a modernist belief in a direct and unproblematic access to clear and certain physical ideas, and a postmodernist indulgence in the notion of an à la carte physics. The intertwining of theory and experiment, inextricably linked by the need to interpret experimental data, does indeed imply that there is an unavoidable degree of circularity involved in scientific reasoning. This means that the nature of science is something more subtle and rationally delicate than simply ineluctable deduction from unquestionable fact. A degree of intellectual daring is required, which means that ultimately the aspiration to write about the *logic* of scientific discovery proves to be a misplaced ambition.[2] Yet the historical fact of the cumulative advance of scientific understanding implies that the circularity involved is benign and not vicious. In assessing the character of science and its achievements, we need to be sufficiently tinged with postmodernism to be able to recognise that there is a measure of rational precariousness involved in its interweaving of theory and experiment, but also sufficiently tinged with a modernist expectation of intellectual attainment to be able to do justice to science's actual success. The philosophical position that mediates between modernism

1. For a memoir of this period, see J. C. Polkinghorne, *Rochester Roundabout*, Longman/W. H. Freeman, 1989.

2. K. Popper, *The Logic of Scientific Discovery*, Hutchinson, 1959.

and postmodernism is commonly called critical realism, the adjective acknowledging the need to recognise that something is involved that is more subtle than encounter with unproblematic objectivity, while the noun signifies the nature of the understanding that it actually proves possible to attain.[3]

I believe that the philosopher of science who has most helpfully struck this balance has been Michael Polanyi. He knew science from the inside, since he was a distinguished physical chemist before he turned to philosophy. In the preface to his seminal book *Personal Knowledge*, Polanyi wrote,

> Comprehension is neither an arbitrary act nor a passive experience, but a responsible act claiming universal validity. Such knowing is indeed *objective* in the sense of establishing contact with a hidden reality . . . Personal knowledge is an intellectual commitment, and as such inherently hazardous. Only affirmations that could be false can be said to convey objective knowledge of this kind . . . Throughout the book I have tried to make this situation apparent. I have shown that into every act of knowing there enters a passionate contribution of the person knowing what is known, and this coefficient is no imperfection but a vital component in his knowledge.[4]

These convictions are worked out in great detail in the book, drawing on Polanyi's scientific experience in a way that other scientists can readily recognise as being authentic. He stresses not only commitment but also the tacit skills involved (for ex-

3. For more extended treatments of critical realism, see J. C. Polkinghorne, *One World*, SPCK/Princeton University Press, 1986, chs 1–3; *Reason and Reality*, SPCK/Trinity Press International, 1991, chs 1 and 2; *Beyond Science*, Cambridge University Press, 1996, ch. 2; *Belief in God in an Age of Science*, Yale University Press, 1998, chs 2 and 5.

4. M. Polanyi, *Personal Knowledge*, Routledge and Kegan Paul, 1958, pp. vii–viii.

ample, in evaluating the adequacy of theories and in assessing the validity of experiments in actually measuring what they are claimed to measure and not some spurious side-effects) that call for acts of judgement that cannot be reduced to following the rules of a specifiable protocol. The method of science has to be learnt through apprenticeship to the practice of a truth-seeking community, rather than by reading a manual of technique, for, in a phrase that Polanyi often repeated, 'we know more than we can tell'. This role for skilful judgement gives scientific research a degree of kinship with other human skilful activities, such as riding a bicycle or judging wine, that require the exercise of similarly tacit abilities. Though the subject of science is an impersonal view of the physical world, its pursuit is an activity of persons that could never be delegated simply to the working of a well-programmed computer. These personal acts of discovery are then offered for assessment and sifting within a competent community, whose judgements are made with the universal intent of gaining reliable knowledge of the physical world. The actual character of our encounter with that world remains the controlling factor. These last points save the personal knowledge of science from fragmenting into a loose collection of individual opinions.

I want to add a further note of a theological kind to the discussion, with the intent of making more intelligible this remarkable ability of scientists to gain such reliable knowledge of the universe, despite the degree of unavoidable epistemic precariousness involved in the endeavour. It is a fact of experience that this repeatedly proves possible, even for phenomena occurring in regimes that are remote from direct human encounter and whose understanding calls for ways of thought quite different from those of everyday life (quantum theory;

cosmology)—a fact, incidentally, that undermines the invocation of Darwinian evolutionary process as an all-sufficient explanation. The widespread success of science is too significant an issue to be treated as if it were a happy accident that we are free to enjoy without enquiring more deeply into why this is the case. Critical realist achievements of this kind cannot be a matter of logical generality, something that one would expect to be attainable in all possible worlds. Rather, they are an experientially confirmed aspect of the particularity of the world in which we live and of the kind of beings that we are. Achieving scientific success is a specific ability possessed by humankind, exercised in the kind of universe that we inhabit. I believe that a full understanding of this remarkable human capacity for scientific discovery ultimately requires the insight that our power in this respect is the gift of the universe's Creator who, in that ancient and powerful phrase, has made humanity in the image of God (Genesis 1:26–27). Through the exercise of this gift, those working in fundamental physics are able to discern a world of deep and beautiful order—a universe shot through with signs of mind. I believe that it is indeed the Mind of that world's Creator that is perceived in this way. Science is possible because the universe is a divine creation.[5]

In its turn, theology is not unacquainted with the necessity of circularity. Augustine and Anselm both emphasised the pattern of 'believing in order to understand' as well as 'understanding in order to believe'. No quest for truth can escape from the necessity of this hermeneutic circle, linking the encounter with reality to an interpretative point of view, so that they are joined in a relationship of mutual illumination and

5. Polkinghorne, *Belief in God*, ch. 1.

correction. Religious insight is not derived from the unhesitating acceptance of fideistic assertion (as if belief were simply imposed by some unchallengeable external authority, conveying to us indubitable propositions), but neither can it be based simply on argument controlled by the conventions of secular thought (such as, for example, the assumption of a purely naturalistic historicism that what usually happens is what always happens). Theology, as much as science, must appeal to motivated belief arising from interpreted experience. Of course, in the case of theology the kind of experience, and the kinds of motivated beliefs that arise from its interpretation, are very different from those appropriate to the natural sciences. The latter enjoy possession of the secret weapon of experiment, the ability to put matters to the test, if necessary through repeated investigation of essentially the same set of impersonal circumstances. This enables science thoroughly to investigate a physical regime defined by a definite scale (such as a given energy range) and to make an accurate map of it. From this ability arises much of the cumulative character of scientific understanding, a linear process in which knowledge increases monotonically. Even in sciences such as palaeontology, where scale is not a controlling factor and significant past events are not repeatable, evidence accumulates in forms that remain permanently accessible, to which direct recourse can be made for further assessment if required.

By way of contrast, in all forms of subjective experience —whether aesthetic enjoyment, acts of moral decision, loving human relationships, or the transpersonal encounter with the sacred reality of God—events are unique and unrepeatable, and their valid interpretation depends ultimately upon a trusting acceptance rather than a testing analysis. The pattern of understanding that results is, so to speak, multidimensional rather than linear, with no necessary implication of a simple temporally ordered increase, as if the insights of the present were inevitably superior in all respects to the insights of the past. Four distinctive features of religious experience express the contrast between science and theology in these respects.

First, there is the fact, already noted, that the development of theological understanding is a more complex process than is the case for scientific understanding. Science achieves cumulative success, accessible in the present without a continual need to return to the past, so that a physicist today understands much more about the universe than Sir Isaac Newton ever did, simply by living three centuries later than that great genius. In religion, however, each generation not only has to acquire theological insight of its own and in its own way, but it also needs to be in a continuing active dialogue with the generations that have preceded it, lest the specific insights that they attained should be lost. In particular, the adherents of a faith tradition have to remain in permanent contact with that tradition's unique foundational events. While contemporary theologians enjoy the opportunities provided by the particular perspective of today, they need also to seek to correct any distortion produced by that perspective by being willing to learn from the complementary insights of earlier generations. All forms of encounter with deeply personal aspects of reality have to take this historical dimension seriously, for the character of their understanding is not simply cumulative, and evaluations need to be made in a living relationship with the past. Just as there is no presumptive superiority of twenty-first-century music over the music of past centuries, so there is no necessary superiority in every respect of the ideas of the theologians of today over those of the fourth or sixteenth centuries. Just as philosophical dialogue today continues to engage with the ideas of Plato and Aristotle, in a similar way the great figures of the theological past—people such as Augustine, Aquinas, Calvin and Luther—remain necessary participants in contemporary conversation, in a way that Galileo, Newton and Maxwell are not so directly involved in the discourse of science. This is because the principal insights of those great scientific pioneers have been incorporated uncontroversially into present textbook knowledge. A purely contemporary judgement suffices. In the case of theology, however, the competent community within which insights are to be received and assessed is not simply the contemporary academy but it is the Church spread across the centuries. Hence the role of tradition, not as a straitjacket imposed a priori on current thinking, but as the indispensable resource for access to a reservoir of attained understanding which has continuing significance.

Second, in placing the physical world under scrutiny, whether by experiment or, in the case of historical sciences such as cosmology and evolutionary biology, by observation, the initiative for setting up this encounter with reality lies with the scientists. In the case of divine reality, however, God can take the initiative in conveying truth and, in fact, all religious traditions believe that this has happened in occasions of revelatory disclosure. One of the prime roles played by sacred scripture in the life of the traditions is to be the record of these theologically foundational events.[6] We have noted already that

6. For more on scripture, see Polkinghorne, *Reason and Reality*, ch. 5; *Science and the Trinity*, SPCK/Yale University Press, 2004, ch. 2.

for the Christian, basic sources of understanding centre on the revelation given in the history of Israel and in the life, death and resurrection of Jesus Christ, a point to which we shall make repeated return.

Making this second point draws our attention to a third difference between theology and science. The motivations for scientific belief arise principally from occurrences that, in principle, are publicly accessible and repeatable, with the consequence that science succeeds in eliciting virtually universal acceptance for its well-winnowed conclusions. Although in its modern form science got going in the particular time and place of seventeenth-century Europe, it has now spread world-wide. Once the dust has settled in some domain of scientific exploration, the insights that have been gained command universal respect and assent. Hence the unanimity, within the relevant competent communities, of belief in the helical structure of DNA and in the quark structure of matter. The religious scene, in contrast, is significantly fragmented. The great faith traditions, such as Judaism, Christianity, Islam, Hinduism and Buddhism, display considerable enduring stability within their adherent communities. They all claim to report and nurture human spiritual engagement with sacred reality, but there is also a perplexing degree of cognitive clash between them concerning detailed belief about the character of these encounters. These disagreements do not relate only to the defining convictions of the religions (such as Christian belief in Jesus as the Son of God, or Muslim belief in the absolute authority of the *Qur'an*), but they extend also to general metaphysical understandings. (Time: a linear pilgrim path, or a samsaric wheel from whose revolutions one needs to seek release? Human nature: qualities uniquely individual and persistent, or recycled through reincarnation?) These clashes seem to exceed anything that could be explained simply as culturally diverse ways of expressing the same underlying truth. In this present book, whose purpose is to explore certain aspects of Christian belief and certain practices of Christian theology, I am not able to do more than acknowledge the challenging and perplexing nature of these interfaith disagreements. Their investigation is an increasingly important and active item on the theological agenda, but one that cannot be pursued further on this occasion.[7]

A fourth point of difference between theology and science relates to the consequences flowing from the embrace of belief. I am entirely convinced of the existence of quarks and gluons, but that belief, illuminating though it is in the limited sphere of elementary particle physics, does not affect my life in any significant way outside the pursuit of intellectual satisfaction in the study or the laboratory. In contrast, my belief that Jesus Christ is the incarnate Son of God has consequences for all aspects of my life, as much in relation to conduct as to understanding. Religious belief is much more demanding than scientific belief—more costly and more 'dangerous', one might say. This means that existential factors play a significant role in the way in which people approach the possibility of religious belief. The motivations that influence its acceptance will rightly include an assessment of the authentic humanity and life-enhancing influences to be found in the believing com-

7. For more on interfaith issues, see the tetralogy: K. Ward, *Religion and Revelation*, *Religion and Creation*, *Religion and Human Nature*, and *Religion and Community*, Oxford University Press, 1994, 1996, 1998, and 2000; also J. C. Polkinghorne, *Science and Christian Belief/The Faith of a Physicist*, SPCK/ Fortress, 1994/1996, ch. 10.

munity. No one could suppose that making this assessment is an unambiguous and straightforward matter. The faiths have all at times been sources of conflict and oppression (as, of course, notoriously has also been the case for atheistic beliefs, such as those expressed in the regimes of Hitler and Stalin). Yet the faiths have also been sources of much human flourishing and centres of compassionate concern for the needy. In the case of Christianity, the dreadful history of crusades and inquisitions has to be held in tension with the Church's record in pioneering education and healthcare, the inspiration and support that it has given to so much achievement in art and music, and the work of many Christian people for peace and justice. There have certainly been ostensibly religious people whose lives have been denials of the values of the gospel, but there have also been many Christians whose lives have displayed outstanding integrity and love. We need to be thankful for St Francis as well as rightly being ashamed of Torquemada.

These four points of difference imply that the defence of critical realism is a more subtle matter in theology than it is in science. The diachronic character of theology, with insight spread across the centuries, deprives it of the simple appeal to manifest monotonic increase in understanding that is so persuasive in the case of science. Yet development and revision certainly occur in theology, as will be demonstrated by some of the illustrative material surveyed in the chapters that follow. Theologians seek to submit their thinking about the divine nature to being shaped by the character of God's revelatory self-disclosures, while acknowledging the ineffable element of mystery present in all human encounter with the Infinite. Thus I believe that theology can rightly lay claim to the pursuit of truth under the rubric of critical realism. Moreover, it can appeal to a theological argument in support of that philosophical claim. The God of truth will not be a deceiver, and insights into the divine character, manifested either in the works of creation or in the events of revelation, can be relied upon not to mislead.

Thus, I see there to be a cousinly relationship between the ways in which theology and science each pursue truth within the proper domains of their interpreted experience. Critical realism is a concept applicable to both, not because there is some kind of entailment from method in one to method in the other—for the differences in their subject material would preclude so simple a connection—but because the idea is deep enough to encompass the character of both these forms of the human search for truthful understanding.

This is a theme that I have often discussed in my writing. Pursuing it requires the analysis of actual examples, rather than relying on an attempted appeal to grand general principles. In my Terry Lectures I sought to set out five points of analogy between two seminal developments, one in physics and one in Christian theology: the exploration of quantum insight and the exploration of Christological insight.[8] In making this comparison, I discerned five points of cousinly relationship between these two great human struggles with the surprising and counterintuitive character of our encounter with reality. In outline, these five points are:

(1) *Moments of enforced radical revision.* The crisis in physics that led eventually to quantum theory began with great

8. Polkinghorne, *Belief in God*, ch. 2.

perplexity about the nature of light. The nineteenth century had shown quite decisively that light possessed wave-like properties. However, at the start of the twentieth century, phenomena were discovered that could only be understood on the basis of accepting the revolutionary ideas of Max Planck and Albert Einstein that treated light as sometimes behaving in a particle-like way, as if it were composed of discrete packets of energy. Yet the notion of a wave/particle duality appeared to be absolutely nonsensical. After all, a wave is spread out and oscillating, while a particle is concentrated and bullet-like. How could anything manifest such contradictory properties? Nevertheless, wave/particle duality was empirically endorsed as a fact of experience, and so some radical rethinking was evidently called for. After much intellectual struggle this eventually led to modern quantum theory.[9]

In the New Testament, the writers knew that when they referred to Jesus they were speaking about someone who had lived a human life in Palestine within living memory. Yet they also found that when they spoke about their experiences of the risen Christ, they were driven to use divine-sounding language about him. For example, Jesus is repeatedly given the title 'Lord', despite the fact that monotheistic Jews associated this title particularly with the one true God of Israel, using it as a substitute for the unutterable divine name in the reading of scripture. Paul can even take verses from the Hebrew Bible that clearly refer to Israel's God and apply them to Jesus (for example, compare Philippians 2:10–11 with Isaiah 45:23, and 1 Corinthians 8:6 with Deuteronomy 6:4). How could this

9. For an introduction to quantum theory, see J. C. Polkinghorne, *Quantum Theory: A Very Short Introduction*, Oxford University Press, 2002.

possibly make sense? After all, Jesus was crucified and Jews saw this form of execution as being a sign of divine rejection, since Deuteronomy (21:23) proclaims a curse on anyone hung on a tree. Experience and understanding seemed as much at odds here as they did in the case of the physicists' thinking about light.

(2) *A period of unresolved confusion.* From 1900 to 1925, the physicists had to live with the paradox of wave/particle duality unresolved. Various techniques for making the best of a baffling situation were invented, by Niels Bohr and others, but these expedients were no more than patches clapped onto the broken edifice of Newtonian physics, rather than amounting to the construction of a grand new quantum building. It was intellectually all very messy, and many physicists at the time simply averted their eyes and got on with the less troubling task of tackling detailed questions that were free from such fundamental difficulties. Problem-solving in normal science is often a more comfortable pursuit than wrestling with perplexities in revolutionary science.

In the New Testament, the tension between human and divine language used about Jesus is simply there, without any systematic theological attempt being made to resolve the matter. It seems that those early generations of Christians were so overwhelmed by the new thing that they believed that God had done in Christ, that its authenticity and power were of themselves sufficient to sustain them without forcing them to attempt an overarching theoretical account. Yet, the position taken by those New Testament writers was clearly intellectually unstable, and the issue could not be ignored indefinitely.

(3) *New synthesis and understanding.* In the case of physics, new insight came with startling suddenness through the theoretical discoveries of Werner Heisenberg and Erwin Schrödinger, made in those amazing years, 1925–26. An internally consistent theory was brought to birth, which required the adoption of novel and unanticipated ways of thought. Paul Dirac emphasised that the formal basis of quantum theory lay in what he called the superposition principle. This asserts that there are quantum states that are formed by adding together, in a mathematically well-defined way, physical possibilities that Newtonian physics and commonsense would hold to be absolutely incapable of mixing with each other. For example, an electron can be in a state that is a mixture of 'here' and 'there', a combination that reflects the fuzzy unpicturability of the quantum world and which also leads to a probabilistic interpretation, since a 50–50 mixture of these possibilities is found to imply that, if a number of measurements of position are actually made on electrons in this state, half the time the electron will be found 'here' and half the time 'there'. This counterintuitive principle just had to be accepted as an article of quantum faith. Richard Feynman introduced his lectures on quantum mechanics by talking about the two-slits experiment (a striking example of counterintuitive quantum ambidexterity), concerning which he wrote,

> Because atomic behaviour is so unlike ordinary experience, it is very difficult to get used to, and it appears peculiar and mysterious to everyone . . . we shall tackle immediately the basic element of the mysterious behaviour in its most strange form. We choose to examine a phenomenon which is impossible, *absolutely* impossible, to explain in any classical way, and which has in it the heart of

> quantum mechanics. In reality it contains the *only* mystery. We cannot make the mystery go away by 'explaining' how it works. We will just *tell* you how it works.[10]

The quest for a deeper understanding of the fundamental phenomena recorded in the New Testament, eventually led the Church to a trinitarian understanding of the nature of God (Councils of Nicaea, 325, and Constantinople, 381) and to an incarnational understanding of two natures, human and divine, present in the one person of Christ (Chalcedon, 451). These were important Christian clarifications, but one cannot claim that theology, wrestling with its profound problem of understanding the divine, has been as successful as science has been in attaining its understanding of the physical world. The latter is at our disposal to interrogate and put to the experimental test, but the encounter with God takes place on different terms, involving awe and worship and obedience. There is an important qualifying theological insight, called apophatic theology, stressing the otherness of God and the necessary human limitation in being able to speak adequately of the mystery of the divine nature. There are bounds to the possibilities of theological explanation. The Fathers of the Church, who at the Councils had formulated fundamental Christian insights, would, I believe, have been quite content to echo Feynman's words, 'We will just *tell* you how it works'.

(4) *Continued wrestling with unsolved problems.* Even in science, total success is often elusive. Quantum theory has been brilliantly effective in enabling us to do the sums, and their

10. R. Feynman, *The Feynman Lectures on Physics*, vol. 3, Addison-Wesley, 1965, p. 7.

answers have proved to be in extremely impressive agreement with experimental results. However, some significant interpretative issues still remain matters of uncertainty and dispute. Chief among these is the so-called measurement problem. How does it come about that a *particular* result is obtained on a particular occasion of measurement, so that the electron is found to be 'here' this time, rather than 'there'? It is embarrassing for a physicist to have to admit that currently there is no wholly satisfactory or universally accepted answer to that entirely reasonable question. Quantum physics has had to be content for eighty years to live with the uncomfortable fact that not all its problems have yielded to solution. There are still matters that we do not fully understand.

Theology also has had to be content with a partial degree of understanding. Trinitarian terminology, for example in its attempt to discriminate the divine Persons in terms of a distinction between begetting and procession, can sometimes seem to be involved in trying to speak what is ineffable. The definition of Chalcedon, asserting that in Christ there are two natures 'without confusion, without change, without division, without separation', is more a statement of criteria to be satisfied if Christological discourse is to prove adequate to the experience preserved in scripture and continued within the Church's tradition, than the articulation of a fully developed Christological theory. Chalcedon maps out the enclosure within which it believes that orthodox Christian thinking should be contained, but it does not formulate the precise form that thinking has to take. In fact, further Christological argument, both within the Chalcedonian bounds and outside them, has continued down the centuries since 451.

(5) *Deeper implications.* A persuasive argument for a critical realist position lies in its offering an explanation of how further successful explanations can arise from a theory, often concerning phenomena not explicitly considered, or even known, when the original ideas were formulated. Such persistent fruitfulness encourages the belief that one is indeed 'on to something', and that a verisimilitudinous account has been attained. In the case of quantum theory, a number of successes of this kind have come to light, including explaining the stability of atoms (their remaining unmodified by the numerous low-energy collisions to which they are subjected), and the very detailed calculations of their spectral properties that have proved to be in impressive agreement with experimental measurements. Strikingly novel, and eventually experimentally verified, predictions have also been made. One of the most outstanding of these is the so-called EPR effect, a counterintuitive togetherness-in-separation that implies that two quantum entities that have interacted with each other remain mutually entangled, however far they may subsequently separate in space. Effectively, they remain a single system, for acting on the one 'here' will produce an immediate effect on its distant partner.

Incarnational belief has offered theology some analogous degree of new insight. For example, Jürgen Moltmann has made powerful use of the concept of divine participation in creaturely suffering through the cross of Christ. He emphasises that the Christian God is the crucified God,[11] the One who is not just a compassionate spectator of the suffering of creatures but a fellow-sharer in the travail of creation. The

11. J. Moltmann, *The Crucified God*, SCM Press, 1974.

concept of a suffering God affords theology some help as it wrestles with its most difficult problem, the evil and suffering present in the world.

The purpose of this book is to pursue further the analogies between the scientific investigation of the physical world and theological exploration of the nature of God. This strategy is adopted in the hope that it will encourage those of a scientific cast of mind to take theological discussion more seriously, and that it will also offer theologians a worked example of a form of possible approach to theological enquiry of a form naturally congenial to the scientifically minded, moving from experience to understanding in a manner that I have called 'bottom-up thinking'.[12] The procedure that I shall follow is to set up a series of parallels between aspects of exploration and conceptual development as we find them respectively in quantum physics and in Christian theology.

12. For an approach to Nicene Christian belief along these lines, see Polkinghorne, *Science and Christian Belief/Faith of a Physicist.*

CHAPTER TWO

Comparative Heuristics

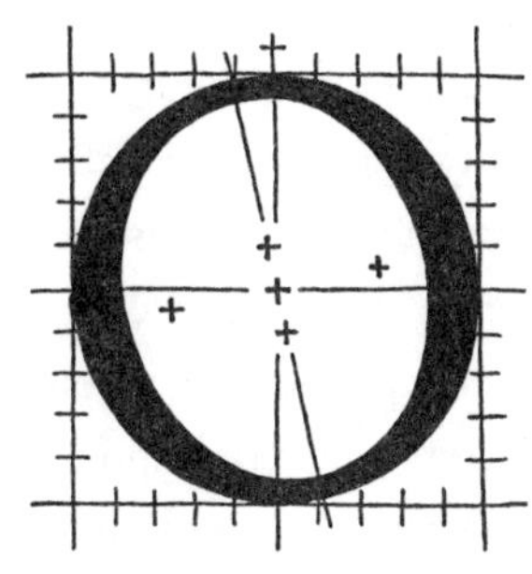

NE of the lessons that science teaches us about physical reality is that its character is frequently surprising. Part of the excitement of doing research lies in the unexpected nature of what may be found lying around the next experimental corner. It is sufficient to utter the words 'quantum theory' to make the point. As a consequence, scientists who are carefully reflective about their activity do not instinctively ask the question 'Is it reasonable?' as if they were confident beforehand what shape rationality had to take. We have noted how 'unreasonable', in classical Newtonian terms, the nature of light turned out to be. Instead, for the scientist the proper phrasing of the truth-seeking question takes the form, 'What makes you think that might be the case?' This style of enquiry is open to the possibility of surprise, while insisting on the presentation of motivating experience to back up any claim that is being made. No unshakable reliance is to be placed on supposed a priori certainties, but evidence is demanded if expectations are to be revised. If you examine what Thomas Young had discovered about diffraction phenomena, and what Albert Einstein had to say about the photoelectric effect, you will be forced to take seriously the seeming paradox of wave/particle duality. In an analogous way, the writers of the New Testament were forced to affirm the even more perplexing fact of their encountering qualities both human and divine in their experience of Jesus Christ.

Neither in physics nor in theology can one remain content simply with accepting the brute fact of the surprising character of reality. There has to be a further struggle to set this new knowledge in some deeper context of understanding. In the case of light, the physicists only began to feel fully at home with wave/particle duality when Paul Dirac's insight led to the discovery of quantum field theory. Because a field is spread out in space and time, it has wave-like properties; when it is quantised, its energy comes in packets that correspond to particle-like behaviour. The threatened paradox had been dispelled by being able to examine a specific example consistently possessing the unexpected dual character. In a similar fashion, Christian thinking could not rest content simply with the assertion of the lordship of Christ, but it had to explore how that lordship related to the fundamental lordship of the God of Israel, a journey of theological exploration that led the Church eventually to trinitarian and incarnational belief.

Considering the ways, both in quantum theory and in theology, in which new discoveries are actually made, and a deeper understanding is then gained, affords a way in which to pursue further the analogies discernible between these two forms of rational enquiry as they engage with their very different kinds of subject material. Similarities emerge in the ways in which experience impacts upon thinking and the manner in which heuristic strategies are developed to yield fuller comprehension. Four exemplary comparisons illustrate the point:

(1) *Techniques of discovery: Experience and understanding.* Advance in understanding requires a subtle and creative interaction between experience and conceptual analysis.

(a) *Theoretical creativity and experimental constraint.* In chapter 1, I stressed the indispensable role played by experiment in driving the development of quantum physics. It is time, now, to redress the balance a little in favour of the theorists by emphasising also the creative role of conceptual exploration. Whatever Einstein meant when he asserted that the fundamental basis of physics had to be freely invented, I am sure that he was not casting doubt on the realistic reliability of our knowledge of the physical world, whose objective existence was a matter of passionate conviction for him. Einstein would never have countenanced any notion of physical theory as an arbitrary intellectual confection. What I think he did have in mind in making the remark was the creative leap of the theoretical imagination that is involved in grasping the character and implications of some great new insight. An outstanding example of this kind of creativity was Einstein's own ability to write down in November 1915 the equations of general relativity, fully formed after years of brooding on the nature of gravity. On the one hand, he must have felt that the economy and elegance of these equations—their possessing that unmistakable character of mathematical beauty that seems always to be present in the successful formulation of fundamental physical theory—was highly persuasive of their validity, but, on the other hand, this did not excuse him from immediately going on to check their consequences against observation. Einstein said that the happiest day of his life was when he found that his new theory of gravity perfectly fitted the behaviour of the planet Mercury, whose motion had long been known to exhibit a small discrepancy with the predictions of Newtonian theory.[1]

The interplay between theory and experiment in physics is deeper than simple dialogue about the interpretation of experimental results. It involves a creative interaction of a profoundly truth-seeking kind between stubborn experimental findings and imaginative theoretical exploration. Truly illuminating discovery far exceeds in subtlety and satisfaction the plodding Baconian accumulation and sifting of a host of particulars, in the hope of stumbling on some useful generalisation. Equally, scientific discovery also exceeds anything attainable simply through free play with mathematical speculation —a point rather sadly made by the second half of Einstein's scientific career, in which he continually and fruitlessly tried to construct general unified field theories, a quest in which the only guide he allowed himself to follow was recourse to ingenious mathematical speculation. Science progresses neither by sole reliance on an earthy empiricism, nor by indulgence in theoretical leaps in the dark, but through the discipline imposed by a continual interaction between assessed experience and proffered interpretation.

1. See, A. Pais, *Subtle Is the Lord . . .* , Oxford University Press, 1982, pp. 250–57.

The way the balance is struck between these two aspects of scientific heuristics changes with varying historical circumstances. I have spoken of how particle physics was largely experimentally driven during my time in the subject. Of course, there were brilliant theoretical insights, discovered by people like Richard Feynman and Murray Gell-Mann, but they arose as responses to empirical challenges. Today, many particle theorists are concentrating on efforts to consolidate and elucidate string theory, and its still somewhat elusive generalisation, M-theory. The fundamental basis for this activity lies in the mathematical exploration of relativistic quantum theory when it is formulated in terms of entities of higher dimensions rather than simple points. Since the combination of relativity and quantum theory is known to yield a synthesis with a depth and fruitfulness that exceeds the mere sum of its component parts, there is good general motivation for such a line of investigation, though there is also such a degree of free creation involved in formulating string theory that it is by no means clear that it will necessarily prove to be the actual description of a new level in the structure of the physical world. At present it lacks the discipline of engagement with new experimental predictions and results. History suggests that quite severe limits should be set on any expectation of human ability to second-guess nature in regimes lying far beyond current experimental access.

(b) *Christology from below and from above.* Scientific progress through a dialectical engagement between experimental challenge and theoretical conceptual exploration has its analogue in theology. An important component in Christological thinking is a careful evaluation of what can be learnt historically about the life of Jesus of Nazareth and about the experiences of the very early Church.[2] Later in this chapter I shall focus on the claim by those first Christians to have been witnesses to encounters with the risen Christ, taking place after Jesus' resurrection from the dead on the third day following his execution. These first-century events, historically unique in their character, are the experiential counterparts for theology of the experiments that initiated the development of quantum physics. Theologians call arguments of this kind 'Christology from below', since the movement of thought is upwards from events to understanding. We shall have much more to say about a theological strategy of bottom-up thinking as the discussion of this book develops. For the present, it suffices to note the analogy that consideration of the historical Jesus bears to the observationally driven component in scientific discovery, and the fact that taking these issues seriously illustrates the point that theological understanding has the character of seeking truth through *motivated* belief, and it is not based on mere fideistic assertion.

Yet, just as physics has to combine experimental challenge with conceptual exploration, so theology has also to complement Christological argument from below with further argument 'from above', assessing the conceptual coherence of the ideas that have been formulated, when they are analysed on their own proper terms. For theology, the tools for this investigation will be provided from the resources of philosophy, in contrast to physics' recourse to the equations of mathematics. But just as not all mathematics proves physically useful, so theology is entitled to discriminate between

2. For introductions to the vast literature, see G. Stanton, *The Gospels and Jesus*, Oxford University Press, 1989; G. Theissen and A. Metz, *The Historical Jesus*, SCM Press, 1998.

the philosophical concepts that it finds helpful for its purposes and those which it has to discard. Philosophical theology has its own protocols and it cannot operate properly using schemes of thought that do not respect its intrinsic character, any more than quantum physics could operate with the criteria of Newtonian thinking. In particular, I believe that an adequate assessment of the historical Jesus cannot proceed under the regulative regime of a naturalistic historicism, whose basic premise is that what usually happens is what always happens.[3] If the physical world often proves to be surprising, so it may also be the case with the revelatory ways in which God has chosen to make known to humanity the divine nature and purposes. Although the warnings of apophatic theology carry the implication that theological discovery is likely to prove less successful and complete than scientific discovery, one can at least see a degree of kinship between the two.

(2) *Defining the problem: Critical questions.* A sharp and selective focus on issues of critical significance is essential to achieve progress in understanding.

(a) *Quark theory.* One of the greatest gifts that a scientist can possess is that of being able to ask the right questions. This ability requires a sense of what is significant and attainable at a particular time in the development of a subject. The discovery of the Standard Model of quark theory proceeded through the successive identification of two key issues that had to be settled. The first arose from the search for an underlying order hoped to be present in the welter

3. See, J. C. Polkinghorne, *Exploring Reality*, SPCK/Yale University Press, 2005, ch. 4.

of new 'elementary' particles that were discovered by the experimentalists from the 1950s onwards. Before the Second World War, Heisenberg had suggested that, since protons and neutrons behave in very similar ways inside nuclei, despite their having quite different electrical properties they might be bracketed together for some purposes and treated as two states of a generic entity that he called a 'nucleon'. The plethora of postwar discoveries of new states of nuclear matter encouraged a greatly enhanced boldness in thinking along these lines. People began to associate particles together in sets, or 'multiplets', whose members had a degree of similarity, while in some other respects seeming quite different from each other. This strategy began to look potentially fruitful when groupings of this kind were recognised as corresponding to certain patterns that mathematically might be associated with structures called Lie groups, and which physically might be thought of as arising from combinations of fractionally charged constituents, the celebrated quarks. A fascinating and suggestive answer had been found to the question of how to introduce some taxonomic order into the particle 'zoo', but this led to the second question of whether this was just a useful mathematical trick, not really much more than an intriguing mnemonic, or whether it was the sign of the presence of an actual underlying physical structure of a quark-like kind. The way to answer this second question turned out to be the investigation of behaviour in an extreme physical regime in which high-energy projectiles bounced off target particles at wide angles. This kind of encounter probed the inner structure of the target in a transparent way. Extremity of circumstance had produced simplicity of analysis. The study of deep inelastic scattering, as these kinds of experiments are called, revealed phenomena that corresponded exactly to the projectiles having struck quarks within the target. In the judgement of the physicists, the reality of quarks had been convincingly established, despite the fact that no single quark has ever been seen in isolation in the laboratory. This character of unseen reality that attaches to the quarks is due, we believe, to a property called 'confinement' that binds them so tightly together within the particles that they compose that no impact is ever sufficiently powerful to dislodge them individually.

(b) *Humanity and divinity.* Theologians, in their turn, need to be clear what are the questions whose answers will control adequate theological thinking in the quest to find an acceptable interpretation of the Church's knowledge and experience of Jesus Christ and its consequent understanding of the nature of God. I believe that there are three such questions of fundamental importance: (i) Was Jesus indeed resurrected on the third day, and if so, why was Jesus, alone among all humanity, raised from the dead within history to live an everlasting life of glory beyond history? (ii) Why did the first Christians feel driven to use divine-sounding language about the man Jesus? (iii) What was the basis for the assurance felt by the first disciples that through the risen Christ they had been given a power that was transforming their lives in a new and unprecedented way?

What are called functional or inspirational Christologies see Jesus as differing only in degree from the rest of humanity. He lived closer to God than others do, and he was more open to the influence of the divine presence that was with him. It is in this fully authentic walk with God that his unique significance is held to lie. Jesus' role is seen as that of providing an example of what a human life might be like, affording encouragement to others to seek to follow that example and conveying to us insights about God, his heavenly Father, that we would not have been able to attain on our own. Those who take this view are sincerely and deeply committed to the cause of Christ as they understand it. They sometimes speak of Jesus in evolutionary terms, seeing him as the first realisation of a newly emergent possibility for human life, an intensification of what had previously been present only partially in the lives of the saints and prophets. According to this view, while Jesus was unique in his time, such a level of life with God might, in principle, be attainable also by others who come later.

I cannot see that this position offers satisfactory answers to the three critical questions of Christology. The resurrection of Jesus, understood in the terms given above and to be discussed in detail later, clearly lies outside anything covered by the idea that he was simply more completely and intensely human than the rest of us. The resurrection could not have amounted to a signal of what to expect in the next stage of hominid evolution, but it must surely have been the result of some great singular act of God. If Jesus was just an unusually inspired man, use of the divine language of lordship about him would seem to have been an unfortunate error, quite inappropriate to someone who was simply a human being, however remarkable. The first Christians testified to the experience of a new power at work in their lives. Such radical transformation of life needs more than example and encouragement, for it requires the gift of divine grace to effect a changed life.

What theologians call the work of Christ—the forgiveness of sins, victory over death, and the bestowal of the Spirit—is an important clue to the nature of Christ. I believe that only an understanding of Jesus that sees in him not only full humanity, but also the fullness of the divine life itself, offers a prospect of meeting adequately the demands made by the New Testament witness to him.

(3) *Expanding horizons: New regimes.* Progress requires allowing novel experience to enlarge the range of conceptual possibility.

(a) *Phase transitions.* New concepts often emerge in physics in response to the exploration of new regimes in which processes are found to display an unexpectedly novel character. The wave/particle duality of light has been our paradigm example, but others could readily be cited. One of the most 'assured' results in physics must have seemed to be Ohm's law, discovered in 1827 and asserting that the current in an electrical circuit is given by dividing the applied voltage by the circuit's resistance. Generations of students had already done countless experiments to verify this prediction when in 1911 a Dutch physicist, Heike Kamerlingh Onnes, showed that, after all, it was not a universal law. When certain metals are cooled to very low temperatures, it is found that their electrical resistance vanishes and a current can circulate without a sustaining electromotive force driving it. Kamerlingh Onnes had discovered superconductivity, a feat for which he was rightly awarded the Nobel Prize in 1913. At the time, no one had the slightest understanding of why this strange behaviour happened. We know now that superconductivity is an intrinsically quantum phenomenon that could not have been understood in 1911. It was only some fifty years later that a theoretical way of accounting for the effect would be discovered.

Of course, the basic laws of physics had not changed at those very low temperatures. They remained the same, but the consequences of these laws altered drastically when one moved from the conducting regime to the superconducting regime. The physicists had had the horizon of their understanding enlarged under the stubborn impact of the strange way metals had proved actually to behave. Transitions of this dramatic kind are called 'phase transitions'. One that is familiar to all of us is the boiling of water. Below 100 degrees Celsius, water is liquid; above that temperature it is gaseous. If we had not seen this transition happen every day of our lives, we would be astonished by it. Phase changes are particularly tricky phenomena to understand, but they illustrate the fact that strikingly different superficial appearances can be consistent with a deep underlying uniformity of basic physical law.

(b) *Miracles.* A rather similar approach is needed in theology in relation to the question of miracles. It does not make theological sense to suppose that God is a kind of show-off celestial conjurer, capriciously using divine power today to do something that God did not think of doing yesterday and won't be bothered to do tomorrow. There must be a deep underlying consistency in divine action, but that requirement does not condemn the deity never to do anything radically new and unexpected. In the Christian tradition, we use personal language about God, not because we think God is an old man with a beard sitting high above the bright blue sky, but because it is less misleading in using the finite resources of human language to call God 'Father' than it would be to employ the impersonal language of 'Force'. The divine consistency is not a rigidly unalterable regularity like that of the force of gravity, but it lies in the continuity of a perfectly appropriate relationship to prevailing circumstances. When those circumstances change radically—when history enters the phase of a new regime, one might say—it is a coherent possibility that that new regime will be accompanied by novel providential phenomena. Thus the problem of miracle is not strictly a scientific problem, since science speaks only about what is usually the case and it possesses no a priori power to rule out the possibility of unprecedented events in unprecedented circumstances. Nor can it claim that its criteria are solely sufficient to define when a significant change of circumstances has occurred. Rather, miracle is a *theological* problem, requiring the discovery of how one might discern divine consistency underlying a claimed particular event, in a way that is compatible with the absence of similar events on other occasions.

Christianity cannot escape the problem of miracle, since the resurrection of Jesus lies at the heart of its belief. Christology is concerned with evaluating the claim that in Jesus human life and divine life are joined uniquely in the incarnation of the Son of God, thereby realising a new regime in the history of creation. These two issues, resurrection and human/divine duality, are central to the theological agenda of this book. They inextricably intertwine. If Jesus is the Son of God, it is a coherent possibility that his life exhibited new and unprecedented phenomena, even to his being raised from the dead to an unending life of glory. (It is important to recognise that the claim being made about Jesus is quite different from that in stories of people like Lazarus, who were said to have been restored to life but who undoubtedly would eventually die again. Jesus was resurrected, not resuscitated.) If Jesus was raised from the dead in this fashion, then there is surely something *uniquely* significant about him. Once again, there is no escape from engagement with the hermeneutic circle in the human search for knowledge of reality. In theology, as in science, one must seek, by scrupulous enquiry into motivated belief, so to tighten that circle that it can be seen to be benign and not vicious.

The attitude to miracles being taken here corresponds to the way in which John's gospel speaks of them as 'signs' (John 2:11, and so on), events that are windows opening up a more profound perspective into the divine reality than that which can be glimpsed in the course of everyday experience, just as superconductivity opened up a window into the behaviour of electrons in metals, more revealing than the discoveries of Professor Ohm had been able to provide. Claims for the occurrence of miraculous events will have to be evaluated on a case-by-case basis. There can be no general theory to cover the character of unique events, but the refusal to contemplate the possibility of revelatory disclosures of an unprecedented kind would be an unacceptable limitation, imposed arbitrarily on the horizons of religious thought.

(4) *Critical events of particular significance.* Specific phenomena, contrary in nature to previous expectation, can confirm radically new forms of understanding.

(a) *Compton scattering.* Much scientific insight arises from a rolling programme of experimental investigation, and the associated developments in theoretical understanding that accompany it. Progress is often gradual and episodic, fought for step by step. The idea of the isolated critical experiment that settles an issue out of hand, a notion somewhat beloved of popular expositors of science, is sometimes the over-simplification and over-dramatisation of a more painstaking process. Nevertheless, there are some occasions when an important matter does seem to receive definite and irreversible settlement as the consequence of a particular experimental result. In the long history of the investigation into the nature of light, such a critical moment occurred in 1923 in an investigation by Arthur Compton into the scattering of X-rays by matter. His results delivered a final *coup de grâce* to any lingering doubts there might have been about the particle properties of radiation.

What Compton had discovered was that the frequency of X-rays is changed by their being scattered by matter. The scattering was induced by an interaction between the incident radiation and the electrons in the atoms that composed the matter. According to a wave picture, these electrons would vibrate with the frequency of the incoming X-rays, and this excitation would cause them in turn to emit radiation *of the same frequency*. Therefore, on the basis of an understanding framed in terms of classical wave theory, no change of frequency was to be expected. On a particle picture, however, what would have been involved was a kind of 'billiard ball' collision between photons and electrons. In this collision, the incoming photon would lose some of its energy to the struck electron. According to Planck's rule, reduced energy corresponds to reduced frequency, with the result that the outgoing scattered radiation would have its rate of vibration diminished, just as Compton had discovered to be the case. It was straightforward to calculate the effect, and the resulting formula agreed perfectly with the experimental measurements. Compton's work had clinched the case for particle-like behaviour, fully dispelling any lingering doubts.

(b) *The resurrection.* The critical question on which all turns in the case of Christology is the resurrection of Jesus. He had a remarkable public ministry, drawing the crowds, healing the sick, proclaiming the coming of God's kingdom. Then, on that final visit to Jerusalem, it all seemed to collapse and fall apart. His entry into the city, riding on a donkey, had been hailed with the politically dangerous cry, 'Blessed is the coming of the kingdom of our ancestor David!' (Mark 11:10), and this had been followed by the religiously provocative act of the cleansing of the Temple (Mark 11:15–18). As the week went on, the mood of the crowd changed and people either turned away or became openly hostile. The civil and religious authorities, Pilate and Caiaphas, acted to regain control of a potentially dangerous situation. Jesus was swiftly arrested, condemned and led away to crucifixion. This painful and shameful death, reserved by the Romans for slaves and rebels, was seen by devout Jews as a sign of God's rejection, since Deuteronomy (21:23) proclaimed a divine curse on anyone hung on a tree. Out of the darkness of the place of execution, there came the cry of dereliction, 'My God, my God, why have you forsaken me?' (Mark 15:34; Matthew 27:46). On the face of it, the final episode of Jesus' life had been one of utter failure. If that had been the end of his story, not only would it put in question any claim that he might have had to any special significance, but I believe that it would have made it likely that he, someone who left no personally written legacy, would have disappeared from active historical remembrance in the way that people do who are humiliated by being seen to have had pretensions above the sober reality of their status. Yet we have all heard of Jesus, and down the subsequent centuries he has proved to be one of the most influential figures in the history of the world. Any adequate account of him has to be able to explain this remarkable fact. *Something* must have happened to continue the story of Jesus. Whatever it was must have been of a magnitude adequate to explain the transformation that came on his followers, changing that bunch of frightened deserters who ran away when he was arrested, into those who would face the authorities in Jerusalem, only a few weeks later, with the confident proclamation that Jesus was God's chosen Lord and Messiah (Acts 2:22–36). I do not think that so great a transformation could have come about simply through calm recollection and a renewed determination to continue to affirm the teaching of Jesus. All the writers of the New Testament believe that what had happened was the resurrection of Jesus from the dead on the third day after his execution.

This is a staggering assertion. Its truth or falsehood is the issue on which the question of the significance of Jesus pivots. The earliest statement of the claim that we have is given in Paul's first letter to Corinth,

> For I handed on to you as of first importance what I in turn had received: that Christ died for our sins in accordance with the scriptures, and that he was buried, and that he was raised on the third day in accordance with the scriptures, and that he appeared to Cephas, then to the twelve. Then he appeared to more than five hundred brothers and sisters at one time, most of whom are still alive, though some have died. Then he appeared to James, then to all the apostles. Last of all, as to one untimely born, he appeared to me. (1 Corinthians 15:3–8)

This passage, composed in the middle 50s, about twenty to twenty-five years after the crucifixion, is an exceedingly condensed piece of writing, wasting no words. It lays an emphasis on witnesses to the resurrection, accessible at the time of writing since most of them were then still living. When Paul says that he is recalling something that he himself 'in turn had received', the natural implication is that he is referring to what he was taught following his dramatic conversion on the road to Damascus. This would place the original testimony within two or three years of the events to which it refers. The idea of such antiquity receives confirmation from the details of the wording, such as use of the Aramaic 'Cephas' for Peter, and reference to the apostles as 'the Twelve', usages that did not last long in the early Church. However, what the appearance experiences were actually like is not made clear in this spare account. To gain insight into that one has to turn to the four gospels.

The New Testament scholar N. T. Wright comments that 'The resurrection stories are among the oddest stories ever written'.[4] Despite their extraordinary subject matter, in many ways the appearance accounts are very matter-of-fact, with little emphasis on the wonder and astonishment that such events might have been expected to induce in their beholders. The stories are somewhat fragmentary, with comparatively little detailed conversation being recorded between Christ and the witnesses. Their character is enigmatic rather than triumphalistic. Instead of immediate exclamations of welcome and relief from the disciples, there is a recurrent inability to recognise at first who it is that is with them,[5] and even after the moment of disclosure it seems that some kind of odd reserve lingers on ('Now none of the disciples dared to ask him "Who are you?" because they knew it was the Lord'; John 21:12). The stories do not follow either of the patterns that might have

4. N. T. Wright, *The Resurrection of the Son of God*, SPCK, 2003, p. 587.
5. Matthew 28:17; Luke 24:15–16; John 20:15; 21:4.

been expected in first-century writings. They portray neither a person resuscitated and restored again from death to ordinary life (like Lazarus; John 11:44), nor the vision of a figure of light (like Revelation 1:12–16). The accounts in the different gospels are distinctly diverse in character. The authentic text of Mark that is now available to us ends at 16:8, without describing the appearance in Galilee that is twice foretold earlier in that gospel (Mark 14:28; 16:7). In Matthew, Jesus appears to women in Jerusalem, but the main story is of an appearance to a group of disciples on a hillside in Galilee (Matthew 28:16–20). In Luke, everything seems to happen in or near Jerusalem on the first Easter day. There is the story of the couple on the road to Emmaus (Luke 24: 13–32), the briefest mention of an appearance to Peter (v. 34), and a subsequent appearance to the disciples gathered in the city (vv 36–51). Yet, the same author writing in Acts (1:3) speaks of appearances occurring over a forty-day period. In John there are appearances in Jerusalem (ch. 20) and an appearance beside the lake in Galilee (ch. 21).

In addition, outside the gospels there is the appearance to Paul 'as to someone untimely born', three times recounted in Acts (9:3–9; 22:6–11; 26:12–18; see also 1 Corinthians 9:1; Galatians 1:15–16). This last story seems different from the others, not only because it occurs significantly later, but also because it has more of the character of a vision of a figure of light. Nevertheless, Paul himself seems to be able to distinguish this event, which he regards as the ground of his claim to be an apostle (1 Corinthians 9:1), from more ordinary visionary religious experience (cf. 2 Corinthians 12:1–5; see also Acts 18:9).

What is one to make of this bewildering variety, so different from the broadly similar ways in which the four gospels tell us about the preceding events that took place in that last week in Jerusalem? Are we simply faced with a collection of tales made up by different authors in different communities as a vivid way of expressing their conviction that the message of Jesus could, after all, continue beyond his death? I do not think so.

A number of considerations persuade me to take the testimony to the appearances of the risen Christ a great deal more seriously than that. First there is that strangely insistent common theme that there was difficulty in recognising who it was, an element of the tradition perhaps most strikingly expressed in Matthew's frank admission that on that Galilean hillside there were some who still doubted (Matthew 28:17). This theme is expressed in different ways in the different stories and I think that it is extremely unlikely that its consistent presence in the accounts is just an accidental coincidence in a bunch of made-up tales. Rather, I believe it to be an actual historical reminiscence of what these remarkable encounters were like. For reasons that one might speculate about, initially it was not easy to grasp that it was the risen Jesus who was there with the witnesses.

N. T. Wright draws our attention to two other unexpected features of the stories. One is their curious reticence about invoking biblical themes to interpret the experiences. Wright calls this 'The Strange Silence of the Bible in the Stories',[6] contrasting it with the way in which themes from the Hebrew scriptures are freely alluded to elsewhere in the

6. Wright, *Resurrection*, p. 599.

gospels, not least in relation to the crucifixion. Yet this biblical contextualisation strangely fades away when it comes to the accounts of the resurrection appearances. It is as though here there is something so new and unanticipated being displayed through God's action, that it has to be considered in its own light alone. Wright says that this change in the gospels' style of presentation gives the 'actual feeling of a solo flute playing a new melody after the orchestra has fallen silent'.[7] His second and related point is to stress the absence from the stories of interpretative glosses linking them to hopes for a destiny beyond death for those who trust in Jesus. Wright emphasises that 'at no stage do they mention the future hope of Christians',[8] in notable contrast to discussions on the resurrection theme elsewhere in the New Testament, such as in 1 Corinthians 15. Once again, it is as though the gospels are content to give raw data without imposing a wider interpretation on what is recorded.

There does seem to be something unique about the gospel accounts of the appearances. They have a very specific character, corresponding to the unprecedented claim that they are making, and they are not just 'more of the same' in a simple continuation that carries the story of Jesus a bit further. Vincent Taylor suggested that the perplexing variations in locale that we find between the different gospels arose because an emphasis on the value of direct testimony meant that each Christian community chose to preserve the account given by its locally accessible witnesses.[9] The more one con-

7. Ibid., p. 600.

8. Ibid., p. 602.

9. V. Taylor, *The Formation of the Gospel Tradition* (2nd edition), Macmillan, 1953, pp. 59–62.

siders these points, the less likely it seems to be that one is dealing with a collection of pious fictions. The style is so different from what might be expected of a confabulation, a point that receives reinforcement when one reads unconvincing second-century attempts in the apocryphal gospels to add further resurrection stories. Wright considers that first-century authors making it all up as a strategy of apologetic or persuasion would not have produced so idiosyncratic a collection of accounts. 'They would have done a better job.'[10] I believe that the appearance stories demand to be taken very seriously indeed as evidence of the truth of the resurrection of Jesus.

There is a second line of evidence pointing to the resurrection that also has to be considered. This relates to the accounts of the finding of the empty tomb. This story is told in all four gospels in essentially the same terms, with only small differences about the exact names of the women and precisely how early in the morning they made their discovery. Once again there is a striking absence of a triumphalist tone in what is said. In fact, the initial reaction to the emptiness of the tomb is one of fear rather than rejoicing.[11] It needs angelic interpretation before its true significance becomes clear. The gospels certainly do not present the story as an instant, knock-down proof of the resurrection.

The first question to ask in assessing this evidence is whether there was indeed an identifiable tomb in which Jesus was buried. The usual Roman custom was to cast executed felons into a common grave, though there is archaeological evidence demonstrating that there were exceptions to this

10. Wright, *Resurrection*, p. 680.
11. Matthew 28:5; Mark 16:8: Luke 24:5; John 20:11.

practice. The case for believing that Jesus was one of these exceptions is strengthened by the part that is played by Joseph of Arimathea in the story. He is an otherwise unknown figure of no obvious importance in the life of the first Christian community, and there seems to be no reason for assigning him this honourable role other than the fact that he actually performed it. In the controversies that soon developed between the growing Christian movement and contemporary Judaism, conflicts that can be traced back into the first century, it is always common ground between the parties that there was a tomb and that it was empty. Where the differences arise is in relation to the reason for this being so. Was it an act of deceit by the disciples or was it resurrection? It seems clear to me which is the more likely explanation.

It is, of course, true that Paul does not refer explicitly to the empty tomb in his writings, which are earlier than the likely date of the first gospel, Mark. Yet, in that spare account in 1 Corinthians 15, Paul takes the opportunity to state that Jesus was buried, which strongly suggests to me that he knew that there was an identifiable tomb. It also seems inconceivable that a first-century Jew like Paul, whose view of human nature would be the psychosomatic picture of our being animated bodies, would have believed that Jesus was alive (as Paul unquestionably did), but his body lay mouldering in the grave. In this connection, Wright makes an ancillary point when he draws attention to the fact that there is no suggestion anywhere of a secondary burial for Jesus.[12] According to Jewish custom, after a corpse had been in the tomb for about a year, so that it had become reduced to a skeleton, the bones would

12. Wright, *Resurrection*, p. 707.

be taken away and placed in an ossuary for interment elsewhere. This apparently did not happen to Jesus. I believe that was because there were no bones to be dealt with in this way.

Yet the strongest reason for taking the story of the empty tomb seriously is that women are assigned the principal role in it. In the ancient world their testimony would have been regarded as unreliable, not to be trusted in a court of law. Any first-century person making up so strange a story would surely have sought to bolster its credibility by making good reliable men the prime witnesses. I believe that the women are credited with the discovery simply because they actually made it. Perhaps they were given that privilege because, unlike those 'reliable' men, they had not run away when the authorities closed in on Jesus.

Thus there are good evidential reasons for taking with the utmost seriousness the claim that Jesus was raised from the dead, however contrary that belief may be to conventional expectation. The case can be strengthened further by two collateral considerations. First, is the Christian establishment of Sunday as the Lord's Day in place of the Jewish Sabbath, in commemoration of its being the day of his rising. This was a change that would surely have required strong motivation of some kind in the very early Church, whose members were pious Jews. Second, the continuing witness of the Church has always spoken of Jesus as its living Lord in the present, rather than as a revered founder figure in the past.

A unique event of this kind cannot be confirmed with the same degree of certainty that attaches to a repeatable experimental finding like that of Compton scattering. Nevertheless, resurrection belief is well-motivated belief that I personally find persuasive. I recognise that how one weighs the claim being made will also depend on how such an astonishing event might find a coherent place in a wider-ranging assessment of the significance of Jesus and his role in God's purposes. Pursuing that analysis is the subject matter of Christology, and a continuing concern in the chapters that follow.

CHAPTER THREE

Lessons from History

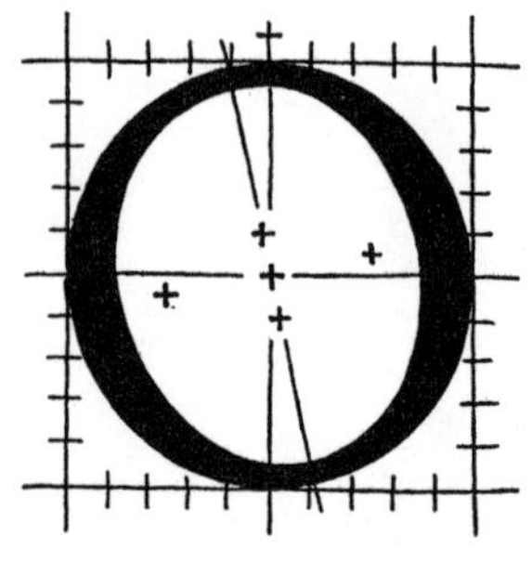

NE of the most influential twentieth-century philosophers of science was Thomas Kuhn. He propounded a celebrated thesis about how science progresses.[1] Most of the time it chugs along, solving problems one by one (Kuhn called this 'normal science'), but once in a while something more dramatic happens. There are times of scientific revolution, moments of radical change in which the scientific 'paradigm' (the currently accepted total view) is drastically altered. An example of such an occasion would be the dawn of quantum theory, with the consequent abandonment of belief in the total adequacy of the classical Newtonian paradigm. A physical world previously considered to be clear and deterministic was found to be cloudy and fitful at its subatomic roots. A theological analogue would be

1. T. Kuhn, *The Structure of Scientific Revolutions* (2nd edition), University of Chicago Press, 1970.

the birth of Christianity, with its radical notions of a crucified Messiah and a risen Lord, ultimately leading to the trinitarian and incarnational paradigm of the nature of God and the relationship between Creator and creatures.

In his enthusiasm for his new idea, Kuhn tended greatly to exaggerate the degree of discontinuity involved in paradigm shift, suggesting that the Newtonians and the quantum theorists inhabited such different theoretical worlds that they could not really talk to each other—despite the fact that a good deal of that kind of exchange actually took place as people tried to understand how measurements are made and how the new physics could secure for itself the actual successes achieved under the old paradigm. Later on, Kuhn himself came to realise that he had overdone it, and he back-tracked somewhat from his earlier extreme position. Nevertheless, he had certainly hit on an important general principle of how to understand what is going on in science, namely that the history of science is the best clue to the philosophy of science. If you want to know how science operates, and what it may legitimately claim to achieve, you have to be willing to study how science is actually done and how its understanding actually develops. Evaluations are to be made on the basis of concrete experience and not by appeal to abstract general principles. The philosophy of science, properly pursued, is largely a bottom-up argument about how things have turned out to be, rather than a top-down argument about how they had to be.

Theology, with its necessarily greater engagement across the centuries with other perspectives beyond that of the present alone (a point that we noted earlier), particularly needs to work with the idea of an historically unfolding development of doctrine. Those who believe in the continuing work of the Holy Spirit (John 16:12–15) will see revelation as a process, rather than a once-for-all act of the communication of instant knowledge. There will certainly be times of special insight, such as the New Testament period, that are comparable to times of revolutionary scientific paradigm shift, but there will also be times of steady 'normal theology'. Our comparative study of science and Christian theology can appropriately turn to considering some historical examples of how the discovery of further truth proceeds in these two disciplines.

(1) *Growing recognition of deeper significance.* Progress in understanding requires a process of the sifting and exploration of the consequences opened up by new conceptual possibilities.

(a) *Theoretical progress.* Planck's original hypothesis, that radiation is emitted and absorbed in packets whose individual energy content is directly proportional to the frequency of the radiation, yielded a splendidly successful formula for the spectrum of black-body radiation. Yet Max Planck himself was far from comfortable with his great discovery, describing it in an interview that he gave later in life, as having been an act of desperation. Quanta are entities that are countable (you have 1,2, 3, . . . packets of energy), and in 1913 Niels Bohr extended this idea of countability to another physical quantity, angular momentum, a measure of a system's rotatory motion. This enabled Bohr to construct his famous model of the hydrogen atom. In quantitative terms, it too was remarkably successful, explaining the properties of the hydrogen spectrum and deriving a formula that had been written down some twenty-eight years earlier by a Swiss schoolmaster,

Johann Jacob Balmer, but which had till then remained simply a numerological curiosity that fitted the facts, although no one knew why. Yet Bohr's principle of the quantisation of angular momentum, in one sense so insightful, was in another sense just an arbitrary imposition of dubious consistency, forced upon Newtonian physics. Not until 1926, when Erwin Schrödinger exploited an analogy with the relationship between wave optics and geometrical optics that enabled him to conjecture his eponymous equation, and then applied this wave equation to the hydrogen atom to find that it yielded a convincing calculation of the Balmer formula, did it seem that real understanding had been gained. A little earlier, in 1925, Werner Heisenberg had discovered the first true formulation of quantum theory, but he expressed it in what at the time seemed to be an unfamiliar and untransparent fashion (matrix mechanics), not immediately recognisable as being the same physical theory that was later proposed by Schrödinger. Events moved fast, however, and the fundamental equivalence of the two theories was soon established, being most clearly demonstrated by Paul Dirac's general account of the principles of quantum mechanics, based on the superposition principle (see p. 18).

The quantum story is one of continuous development within an expanding envelope of understanding. The endpoint (modern quantum theory) looked very different from the starting-points (drops of energy dripping from a black body), yet the pattern had been one of coherent growth in conceptual understanding and effective explanation, linking start to finish and giving the whole episode the character of an increasingly profound grasp of truth.

(b) *Titles and incarnation.* Christian thinking about the

status and significance of Jesus exhibits a corresponding pattern of truth-seeking development. In his lifetime, Jesus was clearly conscious of a specially close relationship to God, his heavenly Father. In Gethsemane he prayed in intimate terms, using an Aramaic word that Mark felt was sufficiently significant for him to need to retain it transliterated in the Greek of his gospel, 'Abba, Father, to you all things are possible, remove this cup from me; yet not what I want but what you want' (Mark 14:36).[2] Earlier, when Jesus had asked his disciples who people said that he was, he had not refused Peter's blurted confession, 'You are the Messiah [Christ]', even though he then told the disciples to be silent about this and he began immediately to qualify the concept of God's anointed one by associating it with suffering and rejection, rather than with military triumph (Mark 8:27–33). The persistence of the title Christ, until it soon effectively became a second name, suggests that the earliest Christians recognised that this was a title of dominical origin that had faithfully to be preserved.[3]

A more complex story attaches to another title that Jesus is portrayed in the gospels as using. The phrase 'the Son of man' is regularly on his lips in all four gospels and only once is it used by anyone else, when a crowd echoes it back to Jesus (John 12:34). Yet there is no reason to believe that the title was in independent regular use in the post-resurrection Church. There has been endless argument among New Testament scholars about what to make of this. Here I can only summar-

2. While there is scholarly argument that *abba* need not necessarily carry an intimate connotation, its occurrence in Mark (and in Romans 8:15 and Galatians 4:6) as a rare Aramaic insertion in the Greek of the New Testament, strongly suggests that it conveys a meaning of special familiarity in these contexts.

3. C. F. D. Moule, *The Origins of Christology*, Cambridge University Press, 1977, pp. 34–35.

ise the conclusions that seem reasonable to me.[4] Sometimes the phrase is clearly referring to Jesus and it implies a special significance attached to him (for example, Mark 8:31), but at other times it refers to a figure related to the final achievement of God's purposes, who is closely associated with Jesus but not unequivocally identified with him (Mark 8:36). I believe that the title goes back to Jesus himself, and the fact that there is some variation in the way in which its relation to him is expressed only seems to confirm this, since it is surely inconceivable that the post-resurrection Christian community would have been in any doubt that Jesus was the final fulfiller of God's purposes. It is very likely that in using the phrase, Jesus had in mind the vision described in Daniel 7, where 'with the clouds of heaven there came one like a son of man' (v. 13, RSV), who is presented before the throne of God to receive 'dominion and glory and kingship' (v. 14). If that is the case, the title carries a very special degree of significance, beyond that of simply referring to a prophetic messenger charged with conveying a word from God.

These titles, Christ and Son of man, tell us something important about Jesus' self-understanding in his lifetime. They are exalted claims, identifying God's special agent in history and beyond history, but they are not intrinsically divine titles. The same can be said of 'Son of God', since this could be attributed to the king of Israel (see Psalm 2:7). To me, these titles are consistent with what a man might think about himself as he lived the truly human life of a first-century Palestinian and believed himself called to fulfil a unique role in God's pur-

4. For a fuller discussion, see J. C. Polkinghorne, *Science and Christian Belief/The Faith of a Physicist*, SPCK/Fortress, 1994/1996, pp. 98–100.

poses, without settling, one way or the other, the question of whether there was more to that man, and to his relationship to God, that might subsequently be recognised by others reflecting on their experience of him, and in particular on their belief in his resurrection.

It is clear that the early Church very soon came to the conclusion that there was indeed more to be said, and from this conviction there arose that strange mixture of divine and human language about Jesus that we have already noted, such as the use of the quasi-divine title 'Lord'.[5] In the search for adequate words in which to speak of Jesus, those early Christians combed the Hebrew Bible to find the resources that they needed. The Roman Catholic New Testament scholar Raymond Brown makes an interesting point when he compares this activity to the way in which the Dead Sea Scrolls community of Qumran handled the Hebrew scriptures. Their technique, called *pesher*, took a prophecy, such as Habbakuk, and sought to interpret it painstakingly, line by line, as referring in detail to contemporary events. The early Christians were much more free and eclectic, their thinking being driven by a concentration on Jesus himself. Brown comments, 'the idea is not primarily that the OT makes sense of the present situation, but that the present situation makes sense of the OT: the control is supplied by Jesus, not by the scriptures'.[6] Among the Old Testament images that the first Christians used were the notion of Jesus as the second Adam who reverses the consequences of the first Adam's fall (see Romans 5:12–21) and the

5. The word *kyrios* had an everyday meaning somewhat similar to the English usage of 'sir', but its use as a manifest title is surely a different matter.

6. R. E. Brown, *An Introduction to New Testament Christology*, Geoffrey Chapman, 1994, p. 98.

figure of divine Wisdom (see 1 Corinthians 1:24). By the time that the prologue to John was written (probably towards the end of the first century) Jesus is identified with the Word—blending, I believe, a concept of the ordering principle of the universe, which is one of the meanings of the Greek word *logos*, with the Hebrew concept of *dabar*, the word of God active in history—and the Word is identified with divinity (John 1:1, 14). Throughout the gospel of John, Jesus is referred to as the Son (in such a manner as inevitably to cause it to be spelt in English with a capital letter) sent by the Father, yet that gospel also contains a verse (John 14:28, 'because the Father is greater than I') to which the Arians would later appeal as supporting their subordination of the Son to the Father. By the end of the New Testament period, there was already much Christological development and some Christological confusion. Theological debate continued, leading eventually to the decisions of the great ecumenical Councils that culminated in the doctrine that the man Jesus of Nazareth was the incarnation of the Second Person of the Holy Trinity. As in quantum theory, so in Christian theology, much greater significance came finally to be recognised than had been apparent at the start of the process of searching for truth. Quantum physics needed for its expression new kinds of mathematics beyond that developed by Newton, and the authors of the Nicene creed had to use philosophical words not found in the New Testament itself. The fifth-century definition of Chalcedon certainly looks superficially very different from its first-century starting-point, but I believe it is possible to conclude, as I shall argue in due course, that what was involved in these developments was not the result of rash and unbridled speculation, but the conceptual development, however deeply counterintuitive, of the implications of the initiating evidence.

(2) *Collateral developments: Further examples considered.* The search for understanding requires the development of a portfolio of inter-connected concepts to do justice to the richness of experience.

(a) *Waves.* Exploration of how conceptual understanding grows historically, in a manner that achieves significant change without severing connection with the initial motivating experience, is so important a theme for my purpose that it seems worth considering some further illustration of this happening, drawn from developments related collaterally to the principal concerns of this book. In the case of physics, it is to wave theory that I turn for an instructive example.

The waves that people first thought about were directly perceptible phenomena, such as the waves of the sea and sound waves induced by vibrating strings. In these examples it was clear that an oscillating material medium served as the carrier of the waves. Therefore, in the nineteenth century when James Clerk Maxwell brilliantly used his new understanding of electromagnetic theory to show that light was made up of electromagnetic waves, it was natural to suppose that in this case also there was a material substrate serving as the medium for carrying the relevant waves. Hence the famous postulate of the existence of the luminiferous aether. Because material bodies appeared to move freely through the aether, it had to have a subtle character, but the nature of the waves that it sustained showed that it also had to be extremely stiff. Not surprisingly, contemporary physicists, such as Lord Kelvin and

Maxwell himself, puzzled over this strange combination of properties. They attempted to conceive of mechanical models of a medium with these characteristics, in an effort to test the consistency of assigning to the aether this kind of behaviour. Their baroque constructions of wheels within wheels were surely not intended to be realistic, but they were thought experiments seeking to test the credibility of the assumptions being made. Perplexity deepened further in 1887 when the Michelson-Morley experiment, designed to measure the velocity of the Earth through the aether, yielded a null result.

In 1905, Albert Einstein cut the Gordian knot by his discovery of special relativity, which promptly resulted in a way of thinking that abolished the need for an aether altogether. Electromagnetic waves were recognised to be just that. It was the energy present in the electromagnetic fields themselves that did the waving.

In 1926, when the Schrödinger equation was formulated as a new kind of wave equation, the question once more arose, waves of what? The initial inclination was to suggest waves of matter, but it soon became clear that this would imply so diffuse an account of the electron that it would not be compatible with its localisation when it was actually experimentally observed. It was Max Born who found the answer. The Schrödinger waves are waves of probability, and the corresponding wavefunction is a representation of the potentialities present in the unpicturable quantum state associated with the electron.

This brief history of the wave idea shows both how indispensable a concept it was in theoretical physics and also how its realistic interpretation moved on from a naive objectivity to a much more subtle account, without at any time ceasing to function as a means for describing the way in which the physical world was found actually to behave.

(b) *Spirit.* A somewhat similar history can be traced in the case of the theological idea of spirit. At the beginning of the Hebrew Bible (Genesis 1:2), the Spirit of God hovers over the waters of chaos (or is it a wind from God, blowing over the waters?—the Hebrew could be read either way). Later there are promises of God's pouring out divine spirit (for example, Joel 2:28–29) and of its focused bestowal on the Messiah (the one anointed by God; see Isaiah 61:1). Spirit is often conceived in the Old Testament in terms of a gift of power, sometimes for very specific purposes (see the account of Bezalel and Oholiab being empowered to construct the ark and the tent of meeting in Exodus 31:1–11). In the preaching of John the Baptist a new note was sounded when he spoke of one who was to come and who would baptise with the Holy Spirit (Mark 1:7–8 and parallels). The gospels and the early Church unanimously identified Jesus as fulfilling this prophecy. In Acts (1:8), the Holy Spirit bestows the power needed for bold witness to Christ, and there are many stories of the signs and wonders that accompany this gift. In the foundational event of Pentecost, it is the risen and exalted Jesus who pours out the Spirit (Acts 2:33). A somewhat different account of pneumatological working is given in the Pauline writings. The gifts bestowed by the Spirit are diverse, distributed in different ways to different believers for different purposes (1 Corinthians 12:4–11), but they yield also the single fruit of a Christ-like life (Galatians 5:22–23). Any merely impersonal notion of spirit simply as power is dispelled by Paul's deeply personal account of the Spirit's intercession with deep sighs, uttered in solidarity with the travail of creation (Romans 8:26–27). In Paul it is God who gives the Spirit (1 Corinthians 2:12 and elsewhere), though he can also speak in one and the same verse (Romans 8:9) of the Spirit, and of the Spirit of God, and of the Spirit of Christ. In John, Jesus both promises to send the Advocate (*Paracletos;* John 16:7–8) and also says that the Father will send the Holy Spirit in Jesus' name (John 14:26).

Everything is not sorted out neatly in the New Testament, but it is clear that mostly its witness to the Spirit is framed in personal terms. It took the Church some time to work out how to think about all this. It was not until the fourth century that the concept of the Holy Spirit as divine, and as being the Third Person of the Trinity, finally emerged with a settled clarity and definiteness in the understanding of the Church. Once again one sees movement from a comparatively naive reification (an extra ingredient given to Bezalel) to a profoundly subtle account of a deep and unexpected reality.

(3) *Tides of fashion.* The questions considered significant, and the style of thinking found appropriate to answering them, are influenced by contemporary intellectual and cultural attitudes.

(a) *Relativistic quantum theory.* At any one time in a branch of science, the attention of researchers tends to be concentrated on a fairly narrow front. Partly this is because current circumstances (the tenor of recent discoveries; what is currently accessible with existing experimental facilities and techniques; and the state of theoretical development) indicate what are likely to be the questions that are feasible to address and profitable to pursue. But there is also another effect operating, which arises from what it is that catches the intellectual imagination of the day. Science is not immune from the tides of fashion, and sociological factors are certainly at work in its communities, even if it is a gross exaggeration to suggest that they determine the nature of the conclusions eventually reached. Consideration of what is fashionable will undoubtedly affect the direction and pace of scientific advance, through its influence on what it is perceived to be appropriate to fund and career-enhancing to work on.

I want to illustrate these effects by giving a brief account of the development of relativistic quantum theory. The subject was initiated by Paul Dirac in the later 1920s through his twin discoveries of quantum field theory and of his celebrated equation of the electron (which is rightly engraved on Dirac's memorial tablet in Westminster Abbey). It was soon realised that all relativistic quantum equations need to be treated as field theories. Initial calculations, based on a rather crude approximation, gave promising answers in agreement with experiment at the level of accuracy that was then attainable. This empirical success, coupled with the conceptual clarification of wave/particle duality and Dirac's successful prediction of the existence of antimatter, clearly showed that the quantum field theorists were on to something.

However, when more refined calculations were attempted, instead of yielding the small corrections that were to be expected, they gave nonsensical results, for the answers turned out to be infinite! Something was going badly wrong. As a result, for a while people lost interest and confidence in quantum field theory. Further progress had to await the end of World War II. The postwar generation of physicists then found an ingenious, if somewhat sleight-of-hand, way round the problem. In quantum electrodynamics (the field theory of the interaction of electrons with photons, abbreviated as QED), it was discovered that all the infinities could be isolated in terms that simply contributed to the mass and charge of the electron. If these formally infinite expressions were replaced by the actual finite values of these constants, the resulting calculations were not only free of infinities but they also proved to be in stunning agreement with experiment. (The predictions of QED differ from measured values by less than the relationship of the width of a human hair to the distance between Los Angeles and New York!) This procedure was called 'renormalisation'.

Quantum field theory had regained its popularity, but it did not last. When attempts were made to apply the same techniques to interactions related to nuclear forces, they failed to give satisfactory answers. Physicists began to question the whole field idea. It is based on the supposition that one can describe what is happening by means of a formalism expressed in terms of all points of space and all instants of time, but in the laboratory we only have much more limited access to what is going on. The basic technique used is a scattering experiment, described simply in terms of colliding particles coming in and scattered particles coming out. It was, therefore, proposed that fields should be replaced by a much leaner account, simply linking 'before' to 'after' the scattering interaction. The resulting formalism was called S-matrix theory (S for scattering). Certain mathematical properties of the S-matrix were known to be implied by relativistic quantum mechanics, and it was hoped that these properties would provide the basis for a new theoretical formulation. It looked promising, and a good number of theorists devoted themselves to the task. In the end, however, the theory became so complicated that it simply collapsed under its own weight.

Just about this time, developments began that were to give field theory a new lease of life. Problems with its application had arisen because the forces relevant to nuclear interactions were too strong for the QED techniques to work. A new class of field theories was identified, called gauge theories, in which interactions were found to become weaker as distances decreased. This meant that some of the old techniques could, after all, be used to discuss nuclear matter in certain circumstances. Field theory once again became the place where the young and ambitious theorist would want to be. This phase has continued so far, and all contemporary theories that are favoured in elementary particle physics are gauge field theories.

(b) *The historical Jesus and the Christ of faith.* If a comparatively straightforward subject like physics has its phases of fashion, one might expect this to be even more the case in a discipline like theology, concerned as it is with a more elusive engagement with profound and mysterious aspects of reality. This has indeed proved to be so. The fluctuating fortunes of quantum field theory have their theological counterpart in the fluctuating emphasis placed in Christological thinking on the importance of research into the historical figure of Jesus of Nazareth.

Critical historical study of the gospel material, conducted in a modern manner, had its origin in the later eighteenth century under the influence of the spirit of the Enlightenment. Yet, from the beginning the limitations of a strongly rationalist approach when applied to this unique material were only too apparent. The posthumously published writings of H. S. Reimarus (1778) illustrated the shortcomings of a reductive analysis when he made the highly implausible suggestion that the disciples stole the body of Jesus and concocted a story of his resurrection in order to promote their dead leader into a spiritual redeemer, a deceitful act undertaken in an attempt to disguise the fact (as Reimarus believed) that Jesus had been more concerned with nationalistic issues than with religious matters. An aversion to any suggestion of the miraculous, and a commitment to a flat historicism based on the axiom that what usually happens is what always happens, characterised much of this initial period of modern research into the historical Jesus which, because it represented an unprecedented turning away from a pre-critical reading of the gospels, is often called 'the first quest'. Perhaps the most notable proponent of such an Enlightenment approach was David Friedrich Strauss, whose influential *Life of Jesus* (1835) made extensive use of the category of myth in giving an explanation of the content of the gospels, and in particular their miraculous element, to which Strauss was willing to attribute only symbolic value. If one understands the category of myth aright, as referring to truth so deep that its most powerful means of conveyance is through story rather than by argument, then it is a concept that certainly is not necessarily hostile to orthodox Christian belief. The critical issue is the relation of myth to history. No one today takes the Greek myths as accounts of actual occurrences, but the orthodox Christian claim that I am concerned to explore and defend in this book, asserts that the story of Jesus' life, death and resurrection is an *enacted* myth, a story that is not only true symbolically but also true historically, as the central component in God's revelatory disclosure within history of the divine nature and the divine purposes. Miracles, such as the resurrection, can then be seen as unique and unprecedented events occurring in unique and unprecedented circumstances.

The first quest of the historical Jesus came to an end with the closing of the nineteenth century, when the tides of Christological fashion began to flow in another direction. Martin Kähler wrote an important book (1892), *The So-called Historical Jesus and the Historic Biblical Christ*, drawing a distinction between the Jesus of history and the Christ of faith, holding the latter to be the real subject of Christian belief. Kähler referred to the 'historic, biblical Christ', but he meant by that the figure who is proclaimed in the imagery of scripture, rather than a person whose actual words and deeds are historically accessible through the evidential record of scripture. One might compare this approach to S-matrix theory, for it is a similar paring down of what is considered significant, in order to concentrate on a directly given final result, without making enquiry into the detail of how that result came about. Consequently, interest shifted from Jesus of Nazareth, who sometimes came even to be thought of as a shadowy and obscure individual, to the commanding Christ-figure who was the subject of the Church's preaching in its proclamation of the gospel. Yet it is surely hard to understand how that figure of the preached Christ arose unless there had been something altogether uniquely striking and remarkable about the wandering Galilean known by those disciples who formed the first cohort of proclaimers of the Christian message.

The final *coup de grâce* to the first quest was delivered in 1906 by the publication of Albert Schweitzer's *The Quest of the Historical Jesus.*[7] He showed how the liberal nineteenth-century accounts of the historical Jesus had been formed much more under the influence of the spirit of their times than by taking this first-century Jew seriously on his own terms. The Catholic writer George Tyrrell expressed the same thought in his witty comment on a great Protestant church historian, that 'the Christ that Harnack sees, looking back through nineteen centuries of Catholic darkness, is only the reflection of a liberal Protestant face, seen at the bottom of a deep well'. Schweitzer himself laid heavy emphasis on the apocalyptic expectation, held by Jesus, of the final breaking-in of the Kingdom of God in order to bring an end to history, believing also that when Jesus felt disappointed in these hopes, he tried to force a divine turning of the wheel of history through seeking his own self-immolation. While it is surely true that Jesus had a much more forceful conception of the Kingdom of God than was congenial to nineteenth-century liberal expectations, Schweitzer equally surely sought to swing the pendulum too far by way of correction. Jesus resists reduction to any simplistic category, whether it be ethical teacher or eschatological prophet.

Rudolf Bultmann was perhaps the most influential New Testament scholar active in the first half of the twentieth century, and he continued a deep distrust of history and a sceptical suspicion of claims of the miraculous. In his opinion, the gospels needed demythologisation if they were to be acceptable to persons living in a scientific age. What mattered about

7. The title of the original German publication was *From Reimarus to Wrede.* Wilhelm Wrede was a New Testament scholar whose book about the so-called 'Messianic secret' in Mark (1901) claimed that Jesus had not believed himself to be the Messiah.

Jesus was his teaching and not the details of his life, which Bultmann thought are largely hidden from us. In his opinion, the category that is essential for seeing what the New Testament is actually all about is existential commitment to the figure of Christ, as proclaimed in the *kerygma*, the apostolic preaching.

Yet commitment to a person unanchored in history because so little could reliably be learned about him might well prove to be commitment to an illusion. In my opinion, a positive evaluation of the relationship between the Jesus of history and the Christ of faith lies at the heart of a credible Christology. Christians cannot rest content with a purely symbolic figure. The religion of the incarnation is inescapably concerned with the issue of the degree of actual enactment that can properly be seen to be involved in the origin of its myth. Least of all in a scientific age can we be satisfied with less than a careful investigation into the historically embedded motivations for Christian belief about the unique significance of Jesus. To treat him as a symbolically evocative, but historically unknown, figure is to lose contact with his reality. It is not surprising that in the second half of the twentieth century, Christological fashion changed again and, in my opinion, changed for the better. A 'new quest' was inaugurated in search of the historical Jesus. It continues vigorously today, in its contemporary phase laying great and justified emphasis on the need to take fully into account the context of first-century Judaism within which Jesus' life was lived.[8] Just as quantum physics was driven to seek a more detailed, and con-

8. For my own accounts see, Polkinghorne, *Science and Christian Belief/Faith of a Physicist*, ch. 5; *Exploring Reality*, SPCK/Yale University Press, 2005, ch. 4.

sequently more illuminating understanding than that afforded by the veiled account of S-matrix theory, so Christology had to return to its foundational roots in the life, death and resurrection of Jesus the Messiah.

(4) *The role of genius.* Advance in understanding owes much to the insights of a small number of exceptional people.

(a) *The founders of quantum theory.* A great deal of development in science stems from the labours of the army of honest toilers in research, as experimental opportunities are exploited and theoretical implications explored. Yet this acknowledgement should not lead us to undervalue the importance of the insights attained by a small band of people with exceptional gifts. There is certainly some truth in the 'great person' version of scientific history, even if it is not the whole story. Particularly at times of great change in the understanding of physical process, the discoveries of the geniuses play an outstanding role in driving the subject forward. In the case of quantum theory in those formative years of the mid-1920s, the exceptional insights of Heisenberg, Schrödinger and Dirac laid the foundations of modern quantum theoretical understanding. Each brought the gift of a particular perspective: Heisenberg's concern with spectral properties and his discovery of the uncertainty principle; Schrödinger's creative formulation of the quantum wave picture; Dirac's profound insight into the underlying mathematical structure, based on the superposition principle. Each of these men won for himself a permanent place of honour in the history of physics. Of course, the times were propitious for great scientific discovery, but these physicists of genius were able to seize these opportunities in a remarkable way that has left its impress on all subsequent quantum thinking.

(b) *Apostolic insight.* The writings of the New Testament are quite varied, but they are dominated by the profound insights of three particular authors, Paul, John and the unknown person who wrote the Epistle to the Hebrews. Different themes find emphasis in their writings: Paul's concern with sin and righteousness, law and gospel; John's emphasis on the love of God and his insight that the cross of Jesus is the throne of glory; the picture given by the writer to the Hebrews of the heavenly intercession of the exalted Christ, who nevertheless had 'learned obedience through what he suffered' (Hebrews 5:8). The depth of theological reflection found in these writings has meant that all subsequent generations of theologians have had to engage with them. It is very striking that the first generation of Christians produced three of the greatest figures ever in the history of theological thought. Their brilliant insights have shaped the form of Christian theology in a manner that the believer will see as the result of providential inspiration by the Holy Spirit, guiding the use of individual human gifts. Of course, the times were propitious for great theological discovery, but these theologians of genius were able to seize those opportunities in a remarkable way that has left its impress on all subsequent Christian thinking.

(5) *Living with unresolved perplexities.* While the ultimate aim is a coherent and fully integrated scheme of understanding, it may be necessary to tolerate living with not all problems fully solved.

(a) *Quantum problems.* How does it come about that the apparently reliable Newtonian world of everyday experience emerges from its fitful quantum substrate? More than eighty years after the initial discovery of modern quantum theory, it is embarrassing to have to admit that there is no comprehensive and universally agreed answer to that reasonable question. Three major problems remain unresolved.

One relates to measurement. When an electron which is in a state that is a mixture of 'here' and 'there' is experimentally interrogated about where it is, using the classical measuring apparatus available in the laboratory, on each occasion a definite answer will be obtained. Not always the same answer, of course, for sometimes it will be found 'here' and sometimes 'there'. The theory enables us to calculate with impressive accuracy the probabilities of obtaining these different answers, but it is unable to explain how it comes about that a specific answer is obtained on a specific occasion. Various proposals have been made, which are mutually incompatible, and none is wholly satisfactory.[9]

The two great twentieth-century discoveries in fundamental physics were quantum theory and general relativity. Straightforward efforts to combine the two yielded nonsensical results. The current attempt at remedy is superstring theory. Its ideas are ingenious, but the approach depends upon trying to guess the form of physics in regimes whose scale is more than sixteen orders of magnitude smaller than those of which we have actual experimental knowledge. Many doubt the feasibility of so ambitious a project.

Microscopic quantum theory and macroscopic chaos theory are imperfectly reconciled with each other. The extreme

9. See, for example, J. C. Polkinghorne, *Quantum Theory: A Very Short Introduction*, Oxford University Press, 2002, pp. 44–56.

sensitivity of chaotic systems means that their behaviour soon comes to depend on fine details of circumstance to which the quantum uncertainty principle forbids access. Yet a synthesis of the two theories is frustrated by their mutual incompatibility. Quantum theory has a scale (set by Planck's constant), but the fractal character of chaotic dynamics means that it is scale free. They do not fit together.

No one rejoices at these perplexities in physics, and all physicists hope for their eventual resolution. Meanwhile the subject is not paralysed in its search for understanding. Scientists can live with partial knowledge and a degree of intellectual uncertainty.

(b) *The problem of evil.* The most perplexing problem that theology faces is the problem of evil and suffering. Its nature can be stated concisely enough. If God is both good and almighty, whence come the disease and disaster, the cruelty and neglect that we observe in creation? If God is good, surely these ills would have been eliminated. If God is almighty, there is surely divine power to do so. Following the century of two world wars, the Holocaust and other acts of genocide, and many natural disasters, this problem presses particularly hard on contemporary Christian thinking.

At least a partial answer can be held to lie, not in qualifying divine goodness, but in a careful analysis of what is meant by 'almighty'. The word does not mean that God can do absolutely anything. The rational God cannot decree that 2+2=5, nor can the good God do evil acts. Almighty means that God can do whatever God wills, but God can only will *that which is in accordance with the divine nature.* Christians believe that nature to be love. The God of love could not be a cosmic tyrant, whose creation was simply a divine puppet-theatre manipulated solely by the divine Puppet-Master. The gift of love is always the gift of some kind of due independence to the object of love. Hence the theological claim, coupled with a deep human intuition, that a world of freely choosing beings is a better world than one populated by perfectly programmed automata. That is the case even if free will results sometimes in horrendous consequences, permitted but not positively willed by that world's Creator. One must admit that this last assertion is not one that one can make without a quiver in the voice, but I believe it to be true nevertheless. It affords some insight into the existence of moral evil, the chosen cruelties and neglects of humankind.

But what about physical evil, disease and disaster? These ills might seem to be much more the direct responsibility of the Creator. I have suggested that there is a kind of 'free-process defence', paralleling the free-will defence in relation to moral evil. All parts of the created order are allowed to act according to their varied natures, being themselves and —through the evolutionary exploration of the potency with which the universe has been endowed—making themselves. In a non-magic world (and the world is not magic because its Creator is not a capricious magician), there will be an inevitable shadow side to fruitful process. Genetic mutations will produce new forms of life, but other mutations will induce malignancy. Tectonic plates will enable mineral resources to well up at their edges to replenish the surface of the Earth, but they will also sometimes slip and induce earthquakes and tsunamis.

These are hard answers to deep and difficult questions, but I believe that they are true insights. They suggest that the ills of the world are not gratuitous, something that a Creator who was a little more competent or a little less callous could easily have eliminated. It is not possible to claim that all perplexity is removed, but I think these theological ideas offer some help with the problem of suffering. Christian theology adds a further level of insight, more profound and existentially relevant than the intellectual discussion we have been pursuing so far. The Christian God is not simply a compassionate spectator of the travail of creation, looking down from the invulnerability of a celestial realm onto the sufferings of Earth. Christians believe that in Christ, and particularly in his cross, we see God sharing human life and its bitternesses, even to the point of a shameful and painful death, experienced in darkness and rejection. The Christian God is the crucified God, truly a fellow sufferer who understands. This insight touches the problem of suffering at the deepest level at which it can be met.[10]

10. For more discussion of the problem of evil, see Polkinghorne, *Exploring Reality*, ch. 8.

CHAPTER FOUR

Conceptual Exploration

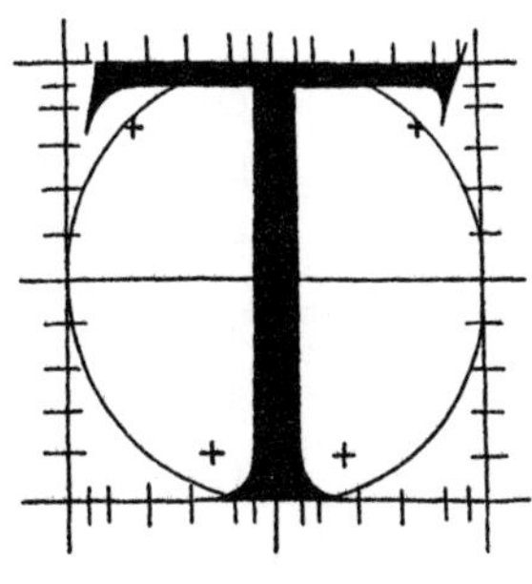

HE bottom-up thinker seeks to move from experience to understanding in the quest for motivated belief. The categories employed in the initial stages of an investigation may not amount to very much more than a rather direct systematisation of the basic evidence, but further reflection can then lead to the formulation of ideas of wider generality and more profound significance, aiming to offer more by way of understanding than simply an immediate matching with the particulars of experience. In this way, deeper insight forms and illumination is gained into the actual character of what is going on. In both science and theology, exploration and clarification of conceptual possibilities are usually attained in an incremental manner.

(1) *Progressive theoretical development.* Continuing conceptual exploration is characterised by increasing subtlety and depth.

(a) *From models to theory.* The first exploration of a new physical regime often takes the form of what the physicists call 'phenomenology', theoretical work formulated in close correlation with specific experimental data. Particular states or processes are investigated and some insight into what is going on can often be gained by the construction of fairly crude models. These models seek to incorporate what appear to be the principal factors controlling the specific phenomena under consideration. They are assembled in what is an admittedly ad hoc manner, simply being put together for the particular purpose in mind. Other factors, not believed to be relevant to the specific issue under investigation, are simply neglected, and there is no pretence that the model thus constructed is a totally adequate description of the nature of the system involved. Models are there to serve strictly limited purposes only.

For example, Einstein's discussion of the photoelectric effect demonstrated the particle-like behaviour of light, without being able to give any account of its wave-like properties. Bohr's model of the hydrogen atom simply imposed an ad hoc rule (the quantisation of angular momentum) on an otherwise Newtonian account. Because models do not aspire to ontological accuracy, it is possible simultaneously to employ a variety of mutually incompatible models in order to gain some purchase on different aspects of the behaviour of the same physical entity. In the early days of nuclear physics, for example, scientists used a model of the nucleus as a cloudy crystal ball in order to interpret results of scattering experiments, a liquid drop model in order to explore nuclear fission, and an independent particle model (roughly treating nuclei in a manner analogous to the discussion of atoms) to understand something about excited nuclear energy levels.

Models are certainly useful in gaining a degree of understanding, but the physicists cannot for long rest content with a portfolio of mutually contradictory pictures of what they are investigating. Some more integrated account has to be sought. A clutch of models needs to be replaced by a single overarching theory, a description with reasonable pretensions to being taken with the ontological seriousness that would correspond to its providing a verisimilitudinous account of what is actually going on, at least at a certain level of detailed structure. Attainment of such a theoretical understanding can come in a variety of ways.

Sometimes new insight comes with astonishing suddenness, as when modern quantum theory sprang into intellectual view fully articulated in those *anni mirabiles* of 1925–26. However, theoretical development is more often piecemeal in its character, passing through a succession of stages. Advances of this kind will unfold in involved and complex ways. Early attempts to find a theory of nuclear properties sought to proceed via the construction of an equivalent potential as a way of describing the forces at work. This was a kind of hybrid approach, a half-way house on the way to a fully quantum account. It made use of a rather conservative strategy, since the idea of a potential is originally a concept belonging to classical physics. An approach of this kind could be expected to have only a provisional significance till something better came along. A more accurate treatment would ultimately require using quantum field theory to describe nuclear processes, an approach that is based on a different idea of interaction altogether, in which forces are generated by the exchange of excitations in the fields, transferring energy and momentum by means of what are called 'virtual particles'. The kinds of fields themselves that need to be considered depend upon the processes that are being investigated. If the energies involved are not too high, it is sufficient to work with fields that correspond to directly experimentally observed particles, such as nucleons and mesons. However, an adequate theory of ultra-high energy interactions would require taking into account more deep-lying structures. For example, very high energy collisions of nuclei are believed, in certain circumstances, to generate a quark–gluon plasma, and attempting to understand this phenomenon requires recourse to the underlying quark field theory (called 'quantum chromodynamics').

Further details are too technical to discuss in a book of this character. However, the essential lesson should be clear. Conceptual exploration proceeds by stages of increasing depth and generality. At each level, contact is maintained with the basic motivating evidence, while at the same time successively greater scope and explanatory power is attained through the novel concepts that are formed. For each aspect of reality being investigated, there is a suitable level of generality and depth that will yield fruitful insight. It would not be an effective strategy to seek to derive the properties of low-level excitations of nuclei directly from quantum chromodynamics, and even less from string theory, if that eventually turns out to be the deeper kind of fundamental account.

While physics is frequently experimentally driven, as in the story of the discovery of the quark level in the structure of matter, sometimes progress comes instead from a deep theoretical re-evaluation of fundamental categories. This re-examination led to Einstein's great discoveries in relativity theory, which sprang from his profound conceptual re-analysis of the character of space, time and gravitation.

(b) *Christological exploration.* Theological thinking, and in particular its Christological engagement with questions of the nature and significance of Jesus Christ, has displayed a similar duality in the way in which progress may be made.[1] There has been Christological argument from below—a discussion which starts from the human life of Jesus, a subject that is then seen to raise issues that point to the need to recognise the presence also of some kind of divine dimension in his life if justice is to be done to the fullness of Christian experience—and there has also been a Christology from above—whose discussion starts with something like the Johannine conviction of the divine Word made flesh (John 1:14) and seeks to explore how the concept of the assumption of human nature by the Second Person of the Trinity might consistently be spoken about. In the early centuries of Christian thinking, the first kind of approach to Christology was particularly associated with thinkers related to Antioch, while the second was particularly associated with thinkers related to Alexandria. It will be clear that, as someone whose intellectual formation has been in science, I feel more at home with the Antiochenes than with the Alexandrians, and it is largely a Christology from below that I want to explore.

Some of the evidential basis on which a Christology from below will have to rest has already been considered. In particular, the issue of the resurrection is clearly central. I agree with Gerald O'Collins that 'the resurrection of the crucified Jesus should be the primary interpretative clue to Christology'.[2]

1. For introductions to Christology, see: R. E. Brown, *An Introduction to New Testament Christology*, Geoffrey Chapman, 1994; G. O'Collins, *Christology*, Oxford University Press, 1995.

2. O'Collins, *Christology*, p. 16.

Yet, while the resurrection surely points to something absolutely unique about Jesus, it does not, in itself, establish his divinity. The New Testament primarily sees the resurrection as a great vindicating act of God, rather than the ultimate miracle performed by Jesus (see verses such as Romans 1:3 and 1 Corinthians 15:4, in the passive voice, but note also that in John (10:18) Jesus is represented as saying 'I have power to lay it [= my life] down and I have power to take it up again. I have received this command from my Father').

The theme of a unique status attaching to Jesus finds confirmation elsewhere in the gospels. Unlike the prophetic figures of the past, Jesus does not simply proclaim a message received from God, but he speaks very directly on his own authority. His characteristic form of address is not 'Thus says the Lord . . .', but 'You have heard that it was said to those of ancient times [that is, by God to Moses at Mount Sinai] . . . but I say to you . . .' (Matthew 5, *passim*). When Jesus heals, he characteristically does so by directly effectual words, rather than by a prayerful invocation of divine assistance. He exercises the divine prerogative to forgive sins in a way that gave offence to some of his contemporaries (Mark 2:5–7).

Two themes strongly present in John are also found, if less insistently expressed, in the other gospels. One is the way in which Jesus lays emphasis on the significance of how people relate to him personally. In John this is most clearly set out in the great sequence of 'I am' sayings ('the bread of life' (6:14); 'the light of the world' (8:12) and so on). The synoptic gospels are more restrained in their language, but the same notion of unique personal significance is present, for example when Jesus says to his disciples, 'whoever welcomes you welcomes me, and whoever welcomes me welcomes not me but the one who sent me' (Mark 9:37 and parallels), or when he speaks in the idiom of heavenly Wisdom, proclaiming, 'Come to me, all you that are weary and are carrying heavy burdens, and I will give you rest' (Matthew 11:28). At the last supper, the bread and wine are said by Jesus to be his body and blood, and the repetition of the rite is to be in remembrance of him (Mark 14:22–25 and parallels; 1 Corinthians 11:23–26). Such words are either extraordinarily self-centred, or they are the truth of the matter.

A second Johannine theme is that Jesus is not simply at the mercy of circumstances, but he is in control of his life through acts of willing acceptance (for example, John 10:17–18). The synoptic counterpart of this is the thrice-repeated prediction of his coming passion (Mark 8:31; 9:31; 10:33–34, and parallels), which is intended to indicate that Jesus knew and accepted what would happen to him when he went to Jerusalem that last time. Many scholars have regarded these predictions as being prophecies after the event inserted by the gospel writers, but I am not inclined to accept that judgement. No doubt the exact wording has been influenced by a subsequent knowledge of the events, but it does not seem at all strange that Jesus should go to Jerusalem that last time keenly aware of an implacable opposition to him on the part of the religious and civil authorities, realising what its consequences were likely to be, and prepared to commit his destiny into the hands of God his Father. I believe that the predictions show us that he accepted the outcome that would follow, and that he trusted in God for his eventual vindication. O'Collins makes a telling point in defence of the passion predictions as stemming from Jesus himself, when he points out that they do not contain any interpretative gloss explaining the events as being 'for us and for our salvation', in the manner one might have expected if they were the creation of the early Church.[3]

There is a final point, made by John Robinson, that one should also note about the earthly life of Jesus. Robinson pointed out that 'the gospels ascribe to Jesus no consciousness of sin or guilt', in striking contrast to that consciousness of sinfulness that has characterised the great Christian saints from Paul onwards (see, for example, Romans 7:21–24). Robinson calls this 'an astonishing omission'.[4] It does not prove the Church's belief that Jesus lived a uniquely sinless life, but it is certainly consistent with it.

The earliest Christians, reflecting on the significance of the life, death and resurrection of Jesus of Nazareth, searched the Hebrew scriptures for interpretative resources to help them make sense of their experiences. The plethora of titles that were associated with Jesus, both in his lifetime and in the very early history of the Church, are the Christological counterparts of the phenomenological use of models in physics. 'Son of man', 'Christ', 'Son of God', 'Lord', 'second Adam', 'Wisdom' are all ways of exploring aspects of foundational Christian experience considered in the light of Old Testament hopes and expectations. They point to Jesus as fulfilling a unique role in God's salvific purposes, but they do not present a coherent and worked-out picture of exactly what this means and how it was accomplished. In addition, there were other aspects of the Church's understanding of Jesus' life and teaching that did not seem to have such clear anticipation in the hopes of Israel. Although the Hebrew scriptures contain

3. Ibid. pp. 71–72.
4. J. A. T. Robinson, *The Human Face of God*, SCM Press, 1972, p. 97.

many references to redemption from sin and evil, to the significance of sacrificial offering, and to the expectation of a new covenant of mercy and peace with God written on the human heart, the New Testament emphasis on the theme of reconciliation (Romans 5:10–11; 2 Corinthians 5:18–20) seems to strike a new note. O'Collins says that reconciliation 'stands practically alone in the NT by not being directly rooted in pre-Christian Judaism'.[5] The Jewish scholar Claude Montefiore felt that he discerned another related element in the teaching of Jesus that did not have a precedent in the witness of the Hebrew prophets. It was the picture of the divine Shepherd who not only welcomes those who turn to God in penitence, but actively goes out into the wilderness to seek the lost sheep (Matthew 18:12–14; but note also Ezekiel 34:11–12).[6] We have already noted what is the most striking discord between the story of Jesus and contemporary Jewish expectations of the Messiah, resulting from his suffering the God-forsaken death of crucifixion. Although Christians soon turned to the figure of the suffering servant portrayed in Isaiah (53:1–12; see also 50:4–9) as a resource for understanding the passion of Christ, this was not a part of general Jewish Messianic expectation.

As in physics, so in Christology, a portfolio of different models fails to satisfy for long and so the more ambitious task soon began of searching for a theoretical setting within which to accommodate these insights in a more coherent unity. In a manner that we have seen also has its analogues in physics, initial attempts at Christological theory-making tended to be somewhat conservative, not departing too far from prior ex-

5. O'Collins, *Christology*, p. 45.
6. Quoted in D. M. Baillie, *God Was in Christ*, Faber, 1961, p. 63.

pectation and thus failing, in fact, to do full justice to the novelty and uniqueness of what was involved. An early attempt corresponded to what later came to be called 'adoptionism'. Jesus was pictured as being a remarkable man fully open to the will of God, who was in consequence adopted at some stage to become the Son of his heavenly Father and so was exalted to the fulfilment of a unique role in the divine plan of salvation. This kind of understanding seems to be expressed in the account in Acts of the first public Christian sermon, given by Peter on the Day of Pentecost. He concludes his address to the crowd by saying,

> This Jesus God raised up, and of that all of us are witnesses. Being therefore exalted to the right hand of God, and having received from the Father the promise of the Holy Spirit, he has poured out this that you both see and hear. . . . Therefore, let the house of Israel know with certainty that God has made him both Lord and Messiah, this Jesus whom you crucified. (Acts 2:32–3, 37)

This speech identifies the resurrection as being the point of adoption, but others came to look to Jesus' endorsement and call by the heavenly voice at his baptism (Mark 1:11 and parallels), or even to his birth. Adoptionism is not without its modern supporters,[7] and it bears a resemblance to an inspirational Christology of a more general kind that simply sees Jesus as a totally Spirit-led human being, different from us in degree but not in any respect different in nature. Yet the main body of Christian thinking soon came to consider adoptionism as inadequate. One difficulty was that it seemed to picture God's relationship with Jesus as being to an extent opportunistic, a

7. For example, J. Knox, *The Humanity and Divinity of Christ*, Cambridge University Press, 1967.

divine trading on the fortunate occurrence of a man found worthy of exaltation. If the life, death and resurrection of Jesus are to have the salvific significance that the New Testament and subsequent Christian experience attribute to them, then all must surely have been from God throughout. The Lukan understanding—that the birth of Jesus was the result of the combination of the divine action of the overshadowing of the Holy Spirit with the human obedience of Mary (Luke 1:30–35)—precisely conveys this message of God's unique initiative in Christ.

The most characteristic New Testament way of expressing God's purposed action in Jesus Christ is in terms of the *sending* of the Son (for example, Mark 9:37; Romans 8:3; Galatians 4:4). An immediate question arises of how to understand this concept, whether it should simply be thought of in terms of a divine pre-determination that there should be such a human person, or whether it requires the much stronger sense of the taking of human life by an already existent being. This latter sense of the sending of one who was pre-existent, seems to be the more natural, if more mysterious, interpretation. Certainly, by the time that John's gospel was written, this way of thinking was being strongly expressed. In John, Jesus repeatedly speaks of himself as the one who has been sent by the Father (4:34; 5:23–24; and so on), and in the Prologue we are told that the divine Word, that was from the beginning, became flesh and lived among us (1:1 and 14). The epistle to the Hebrews (1:1–4) says something rather similar, expressed in its very different idiom. An absolutely central issue in Christology is whether sense can truly be made of the idea of the incarnation of the pre-existent Son, the assumption of finite humanity by the infinite Word.

One early response to the need to assign a unique status to Jesus took a form that, in fact, opposed the idea of his taking true humanity at all. The theory called 'docetism' supposed that Jesus was actually a spiritual being who just appeared to be human. This is the one Christian heresy that soon failed to gain any degree of continuing intellectual respectability, though one must admit that a good deal of popular Christian piety has tended to be somewhat docetic in its reluctance to ascribe to Jesus any of the limitations inescapably associated with a truly human life. He has sometimes been credited with a veiled form of superhuman knowledge, as if this first-century Palestinian must have known about quantum theory. Of course, the Word by whom all things were made must know the true nature of physics, but if that Word really took flesh, that must have involved a self-emptying of divine omniscience in the embrace of human finitude. This latter insight is called 'kenotic Christology', derived from the Greek word for emptying, and it takes much of its inspiration from a passage in Philippians (2:5–11) where Paul speaks of Christ as one 'who, though he was in the form of God, did not regard equality with God as something to be exploited, but emptied (*ekenosen*) himself, taking the form of a slave, being born in human likeness'. In the gospels, Jesus is portrayed as facing the ambiguous conflicts of duty that no human person can wholly avoid, both in respect of family (Mark 3:31–34) and in respect of the scope of his mission (Mark 7:24–30). The Johannine writings, despite their high Christology, are insistent on the reality of the Word made *flesh*, asserting that 'every spirit that confesses that Jesus Christ has come in the flesh is of God, and any spirit that does not [so] confess Jesus is not from God' (1 John 4:2–3).

An important theological issue is at stake here. I suggested earlier that the work of Christ (his bringing salvation and new life to humankind) is an essential clue to the nature of Christ. From the very first it has been recognised that this salvific role depends critically upon Jesus' solidarity with humanity, so that he is truly one of us and so relevant to us, while also requiring in him the unique presence of the divine life that alone is powerful enough to overcome human sin and our alienation from the God who is the true ground of our being. In the second century, Irenaeus put this very clearly when he wrote,

> If a human being had not overcome the enemy of humanity, the enemy would not have been rightly overcome. On the other side, if it had not been God to give us salvation, we would not have received it permanently. If the human being had not been united to God, it would not have been possible to share in incorruptibility. In fact, the Mediator between God and human beings, thanks to his relationship to both, had to bring both to friendship and concord, and bring it about that God should assume humanity and human beings offer themselves to God.[8]

A modern writer on Christology, John Knox, made the same point when he wrote 'How could Christ have saved us if he were not a human being like ourselves? How could a human being like ourselves have saved us?' going on to ask, 'Who could have saved us but God himself? How could even he have saved us except through a human being like ourselves?'.[9] A satisfactory answer to these questions demands some concurrence of divinity and humanity in Jesus.

8. Quoted in O'Collins, *Christology*, p. 155.
9. Knox, *Humanity and Divinity of Christ*, pp. 52, 92.

These theological considerations, together with the general pressure felt by the Church from the earliest times to have recourse to both divine and human language in its attempt to speak adequately of Jesus, constitute the motivations that drove Christian thinking to its continuing exploration of how to conceive of the relationship between Jesus and his heavenly Father. Knox summarized the issues neatly when he wrote, 'the man Jesus most surely remembered and the heavenly Lord most surely known—and the age-old problem of Christology is implicit in that fact'.[10]

There is also a second major theological point at stake, though one whose central significance perhaps only came to be fully appreciated in the twentieth century. In the aftermath of the Holocaust, theologians came to realise with even greater clarity than before that the problem of the world's suffering demands that God should be more than a detached, if compassionate, spectator of the travail of creation. God must also truly be a fellow-sharer in the world's pain. We have already noted how the concept of the crucified God, who in the cross of Christ is caught up as a fellow-participant in suffering, plays so powerful a part in contemporary theological struggle with the perplexity of evil. True divine sharing in the darkness of suffering and death is the deepest possible response to the task of understanding the strangeness of creation.[11] The cross and the resurrection together afford the ultimate affirmation and ground of hope that the last word will be with the God who fully participates in creation's suffering and thereby redeems it, and not with the forces of evil. These profound insights re-

10. Ibid., p. viii.

11. P. Fiddes, *The Creative Suffering of God*, Oxford University Press, 1988.

quire that God was truly 'in Christ' (2 Corinthians 5:19) in a great act of solidarity and rescue. Only a fully incarnational Christology can reach the depth that the divine response to suffering requires.

The next big attempt at Christological theorising surfaced in the late third century in an attempt to find a halfway house between divinity and humanity in thinking about the nature of Christ. Arianism conceived of Christ as the first created being, Son of God in a subordinated sense and fulfilling the role of an intermediary between the Creator and the lives of all other creatures. It appealed to texts such as John 14:28 ('the Father is greater than I'), which others interpreted simply as referring to Jesus' kenotic acceptance of the necessary limitations of a truly human life, and Colossians 1:15 ('the first born of all creation'), which those who opposed the Arians regarded as being just a manner of speaking used to emphasis the supreme significance of Christ in relation to the created order (which is certainly the principal theme of the Pauline passage from which the quotation is culled). For a while Arianism made great strides in the Christian world, though always vigorously opposed by Athanasius of Alexandria. In the end, however, the Arians were defeated. The great flaw in their Christological thinking was its breaking the salvific link between the life of God and the life of humanity. A kind of demi-god seemed neither to be sufficiently related to humanity to be relevant to human needs, nor sufficiently related to God to bring about the salvation that only truly divine power could accomplish.

The formal theological defeat of the Arians came in 325 at the Council of Nicaea, which affirmed the divinity of Christ by proclaiming him to be 'of one substance' with the Father.

Yet the debates and disagreements continued in the Christian community during much of the rest of the fourth century, until matters were further clarified at a second Council held at Constantinople in 381. The Greek word that had been used to assert the unity in divine essence of the Father and the Son is *homoousios*, meaning of the same substance. It contrasts with the similar-sounding but theologically different word *homoiousios*, meaning of similar substance, which some (called semi-Arians) would have preferred as a means of maintaining a clear distinction between the Father and the Son, but which was deemed by the majority to be inadequate. Some were worried by the use of a philosophical term like *homoousios*, which does not appear in the New Testament. The fact is that the theologians had found it necessary to go beyond biblical language, anchored in direct experience, and to adopt a more philosophical tone of discourse if justice were to be done to the profundity of what they were struggling to understand, just as the language of physical theory (quantum fields, strings) has had to go beyond the phenomenological discourse of direct experimental observations (energy levels, multiplets).

While the Creed that stemmed from Nicaea and which was later augmented at Constantinople affirmed the divinity of Christ, it was equally clear and insistent in affirming his true humanity, asserting that he 'was made man'. Both halves of the soteriological duality of divinity and humanity were seen to be essential to adequate Christological thinking. Subsequent discussion of an orthodox kind (that is to say, perceived by the Church to be in accord with fundamental scriptural and ecclesial testimony) has had to wrestle with the paradox of how infinite divinity and finite humanity could be co-present in a single person.

Various theoretical attempts were made to cope with this problem that did not, in the end, seem acceptable to Christian understanding. Apollinarianism suggested that the divine Word had taken the place of the human soul in Christ, but this was found not to maintain adequately the full humanity of Jesus. Nestorian thinking tried to place divinity and humanity side-by-side in an almost independent way, so that when Jesus met the woman at the well in Samaria, his humanity was weary and thirsty, but it was his divinity that discerned her past history (John 4:5–26). This was rejected because so partitioned an account prejudiced the integrity of Christ's person. The followers of Eutyches tried so to mingle the divinity and humanity together that they simply made Jesus into an oddity, a kind of spiritual centaur. In 451, the Council of Chalcedon framed its celebrated definition,

> one and the same Christ, Son, Lord, Only-begotten, recognised in two natures, without confusion, without change, without division, without separation; the distinction of the natures being in no way annulled by the union, but rather the characteristics of each nature being preserved and coming together to form one person and subsistence.

The Chalcedonian formula, two natures in one person, did not in itself solve the Christological problem, but it was a specification of what the Church held would be necessary to find in a solution if it were to be consistent with the Christian experience of Christ. O'Collins summarises the post-Chalcedonian situation in the words,

> In synthesising the concerns and insights of the Alexandrian and Antiochene schools, Chalcedon provided a 'logical' conclusion to the first three ecumenical councils.

> Against Arianism, Nicea used the term *homoousios* to reaffirm 'Christ is divine'. Against Apollinarianism, Constantinople affirmed 'Christ is human'. Against Nestorius, Ephesus [a council held in 431] professed that Christ's two natures (his divine being and his human being) are not separated. Against Eutyches, Chalcedon confessed that, while belonging to one person, the two natures are not merged or confused.[12]

In this way, the area for the containment of an acceptable Christological discourse was marked out by the Fathers. Subsequent orthodox discussion has remained in a state reminiscent of that of the physicists between 1900 and 1925. The latter were forced by their experience to embrace a picture of the wave/particle duality of light, without at that time having a deep physical theory to explain how this might consistently be so. Christian thinkers have been forced by experience to embrace a picture of divine/human duality in Christ, without having yet attained a deep theological theory of how this might be so. The Christological counterpart of quantum field theory still remains to be discovered. Perhaps the warnings of apophatic theology indicate that this must necessarily remain the case, at least in this life. There can be times when one just has to hold on to the strangeness of experience by the skin of one's intellectual teeth, knowing that progress will not come from a facile abandonment of any part of that experience.

Living with unresolved paradox is not a comfortable situation. Yet it is not an unfamiliar state for men and women to find themselves in. Even quantum physics has its unsettled questions, such as the measurement problem. In fact, Niels Bohr once said that anyone who claimed fully to understand

12. O'Collins, *Christology*, p. 194.

quantum physics had just shown that they had not begun to appreciate properly what it is all about. He was echoing, unconsciously no doubt, a similar remark made earlier by William Temple when he said that 'if any man says he understands the relation of Deity to humanity in Christ, he only makes it clear that he does not at all understand what is meant by Incarnation'.[13] On another occasion, Bohr said that there are two kinds of truth, trivialities where to embrace both opposites would obviously be absurd, and profound truth, recognised by the fact that the opposite is also a profound truth. The essential thing about seeming paradox, whether in science or in theology, is that it should be forced upon us by experience and not just embraced in a fit of unrestrained speculative exuberance.

In his book on Christology, Donald Baillie pointed to what he called the 'Central Paradox' of the Christian life. He was referring to the convictions simultaneously held, that we bear a responsibility for our lives and actions, and also that 'never is human life more truly and fully personal, never does the agent feel more profoundly free, than those moments in which he can say as a Christian that whatever was good was not his but God's'.[14] Paul expressed a similar thought when he exhorted the members of the church at Philippi to 'work out your own salvation; for it is God who is at work in you, enabling you both to will and to work for his good pleasure' (Philippians 2:12–13). Baillie went on to suggest that 'this paradox in its fragmentary form in our Christian lives is a reflection of that perfect union of God and man in the Incarnation on which the whole Christian life depends, and may

13. W. Temple, *Christus Veritas*, Macmillan, 1924, p. 139.

14. Baillie, *God Was in Christ*, p. 114.

therefore be our best clue to understanding it'.[15] A scientist should not be unsympathetic to the notion of an appeal to experience being the best strategy to pursue in the search for understanding.

(2) *Indefiniteness: A cloud of unknowing.* No greater clarity should be sought than reality permits.

(a) *Wave/particle duality.* It is worth understanding in a little more detail how quantum field theory reconciles the apparent opposites of wave and particle behaviour. This possibility is found to result from the fact that states corresponding to wave-like properties contain an indefinite number of particles. This is a property that Newtonian physics, of course, could not accommodate, for in its clear and determinate formulation there would simply be a specific number of particles present (just look and count them) and that would be that. In quantum theory, however, the superposition principle allows the addition of possibilities that classical physics would hold strictly apart, so that a state can be composed of a mixture of different particle numbers, with no fixed and definite number present. It is the ontological flexibility of the quantum world, whose description in terms of wavefunctions expresses present potentiality rather than persistent actuality (consequently incorporating an element of intrinsic indefiniteness into its account), that dissolves the paradox of wave/particle duality.

(b) *Christological duality.* Perhaps theology can take heart from this example of quantum thinking. The Church resisted the easy solution of the manifest separations proposed by Apollinarianism and Nestorianism in their different ways, and

15. Ibid., p. 117.

it affirmed the more nuanced and less specifically articulated Chalcedonian formulation. Acceptance of a degree of mysterious indefiniteness is a stance that does not seem totally foreign to the quantum physicist.

(3) *The toys of thought.* Simplified conceptual structures can be used as aids to the exploration of possible schemes of thought.

(a) *Thought experiments.* Physicists sometimes seek to gain insight into the nature of novel concepts by the mental exercise of constructing idealised and highly simplified situations to which they could be applied. The most famous of these 'thought experiments' was a series of encounters between Albert Einstein and Niels Bohr in the early days of modern quantum theory. Einstein was highly distrustful of the way in which quantum thinking had developed, and he tried to think of a series of schematic measurement procedures that would, in principle, contradict Heisenberg's uncertainty principle. His proposals gave Bohr some sleepless nights but, in the end, the latter was always able to show that the apparent violation was due to a failure to apply the restrictions of uncertainty with complete thoroughness. Eventually Einstein had to concede defeat on this issue, but the discourse based on thought experiments had served to produce a useful clarification of quantum principles.

(b) *Pictures of the eschaton.* Christian hope of a destiny beyond death is framed in terms of the eschatological expectation that the present 'old creation' will be transformed into God's 'new creation', a world from which transience and decay will have been banished for ever. Such a world is quite different from that of our present experience, and it is beyond the power of contemporary human thinking to give an adequate description of it. Yet Christian writing, both in the New Testament (for example, Revelation 21:1–4; 22:1–5) and later, has sought to give some symbolic expression of what this hope might be like. The intent, properly understood, is not to give a timetable for the end of history, or a map of heavenly geography, but pictures are offered of the redeemed life of the world to come that can be considered as theological thought experiments, exploring to a modest degree the conceptual content and coherence of Christian hope.[16]

(4) *Major revision.* Under the pressure of new experience or further insight, long-held convictions may require radical revision.

(a) *Physical determinism and indeterminism.* For two centuries, Isaac Newton's mathematisation of physical thinking, expressed in equations whose solutions are uniquely determined by the specification of appropriate initial conditions, had suggested to many people the picture of a clockwork universe of tightly determined process. However, twentieth-century physics saw the death of this kind of merely mechanical view of the world, a consequence brought about by the discovery of intrinsic unpredictabilities present in nature, first at the level of atomic phenomena (quantum theory) and subsequently at the level of everyday phenomena (chaos theory). Unpredictability is an epistemological property concerned with what can be known, and it is a matter for metaphysical discussion and decision how it might be interpreted ontologically, as a description of what is actually the case. Is unpredictability to be understood simply as an unfortu-

16. See, J. C. Polkinghorne, *Exploring Reality*, SPCK/Yale University Press, 2005, ch. 10.

nate matter of inescapable ignorance, or is it the sign of an openness to the influence of causal factors which go beyond the scientific story of energy exchanges between constituents? Physical process has certainly proved to be something more subtle than previous generations of physicists had supposed, and we may choose the metaphysical option of believing it also to be more supple. While controversies continue concerning details of the metascientific significance that should be assigned to these twentieth-century discoveries of unpredictability (for example, there are alternative indeterministic and deterministic interpretations of quantum theory with indistinguishable experimental consequences[17]), science's contribution to forming ideas about the causal structure of reality undoubtedly underwent a radical re-evaluation in the course of that century. One can take with absolute seriousness all that physics actually can tell us, and still believe in a world of true becoming, in which the future is not just an inevitable consequence of the past. A careful evaluation of what science has to say shows that it is not inconsistent with our human experience of the exercise of agency, for physics has not established the causal closure of the world on its own terms alone. Neither can it be used to deny the possibility of divine providential agency.[18]

(b) *God and time.* People sometimes criticise theology as being unwilling to subject its thinking to any form of significant reassessment. While the rate of revision may be less rapid than that in science, it would be a gross mistake to think that

17. See, J. C. Polkinghorne, *Quantum Theory: A Very Short Introduction*, Oxford University Press, 2002, pp. 53–55.

18. J. C. Polkinghorne, *Belief in God in an Age of Science*, Yale University Press, 1998, ch. 3.

theology is frozen into a state of unchanging bondage to the concepts of the past. Recent re-evaluations of how to conceive of God's relationship to time make it clear that this is not so.

Classical theology, stemming in the West from Boethius and Augustine, further developed by Thomas Aquinas in the Middle Ages and adhered to in Reformation times by thinkers such as John Calvin, pictured the divine nature as being wholly atemporal, with God totally outside of time, looking down, so to speak, onto the whole history of creation laid out before the divine gaze *totum simul*, all at once. According to this view, God did not have *fore*knowledge of the actions of free creatures, but simply knowledge, for all moments of created time were supposed to be equally contemporaneous to the eternal deity.

Three considerations have led a number of contemporary theologians to propose a radical revision of this idea, in the direction of envisaging an engagement of God with the reality of time, while not denying the complementary reality of divine eternity. One motivation for this move is the scientific discovery that physics' actual knowledge of the character of process can be interpreted as being consistent with the picture of a world of true becoming, in the manner sketched above. In an open universe, divine interactive providence can be an historically effective part of what brings about the future. If such is the character of the world, since God surely knows things in accordance with their true nature, God will not only know that the events of creation are successive, but they will be known in their succession. This implies a true divine engagement with time. Of course, God is not in thrall to time as creatures are, so that a theology of this open kind takes a dipolar view of the divinity, locating the presence of both eternity and temporality within the divine nature.

The second consideration in favour of dipolarity appeals to the scriptural picture of a God who is deeply involved in the unfolding history of Israel and in the temporal episode of the incarnation, in addition to being the God on whose eternally unchanging steadfast love all creatures can depend. The third consideration relates to the criticism that the thinking of classical theology had laid such strong emphasis on divine transcendence, conceiving God as totally distant from creatures, that it failed to recognise that the God whose nature is love must also be immanently present to those creatures who are the object of that love, even to the point of an engagement with time.[19] Dipolar concepts of the divine relationship to time and eternity are matters for argument in contemporary theological circles, but the vigour of the debate makes it clear that religious thinking is not unwilling to take the risk of conceptual revision.

(5) *Grand Unified Theories.* Much profound human thinking is inspired by a search for understanding based on integrated relationality.

(a) *GUT.* One could write the history of modern physics in terms of its being a continuing quest for greater generality and deeper unity in our conceptual understanding of the physical world. It all began as early as Galileo's conviction, contrary to the thinking of Aristotle, that the heavenly bodies are made of the same materials as those that compose terrestrial entities. Sublunar physics is the same as celestial physics. This insight was triumphantly confirmed by Isaac Newton's

19. See, Polkinghorne, *Exploring Reality*, ch. 6.

discovery of universal gravity, showing that the force that makes the apple fall is the same force that holds the Moon in its orbit around the Earth.

The next unifying steps occurred in the nineteenth century. At first sight electricity and magnetism seemed as different from each other as the action of rubbing a glass rod with cat's fur differs from suspending the lodestone. Yet the experimental discoveries of Hans Christian Øersted and Michael Faraday showed that there was some direct connection between electric currents and magnetic fields. The character of this connection was made clear in 1873 when James Clerk Maxwell published his *Treatise on Electricity and Magnetism*, presenting a unified theory of electromagnetism that has proved of lasting value, and which is one of the most brilliant achievements in the whole history of theoretical physics.

The next great unificatory advance came from a marriage between electromagnetic theory and the weak nuclear forces that are responsible for phenomena such as ß-decay, the emission of electrons by radioactive nuclei. At first sight, this too seemed a highly unlikely coupling, since the weak forces, as their name suggests, are much feebler than electromagnetism and they exhibit properties, such as an intrinsic chirality (a kind of preference for left-handed configurations), that electromagnetic phenomena do not display. Nevertheless, in the late 1960s Steven Weinberg and Abdus Salam independently hit upon a unified electroweak theory that puts the two together in a manner that has proved theoretically consistent and empirically successful.

The next desirable step would obviously be a further integration, drawing in the strong nuclear forces, and perhaps also gravitation (which is by far the weakest of the basic forces of nature). Such a Grand Unified Theory, or GUT for short, has so far proved difficult to achieve and the attempts to find it have been controversial and not wholly convincing. The present favoured candidate is superstring theory, but accepting its ideas depends upon believing that theorists, on the basis of mathematical considerations alone, can second-guess the character of nature at a level of detail more than ten thousand million million times smaller than anything of which we have direct empirical experience. The lessons of history are not encouraging to such a bold venture. Usually nature has something up her sleeve that only empirical pressure will cause the theorists to think of.

Whatever may eventually prove to be the case, the general hope that some form of GUT will in the end be discovered is one that is entertained by many physicists, myself among them. A belief in the fundamental unity of physics is one that is encouraged by the kind of past experiences that we have reviewed. It is also supported by a metaphysical conviction of the integrity of cosmic process that is deeply appealing to scientists. Theologians may well feel that this act of faith by the physicists is a reflection of a trust, doubtless often unconsciously entertained, in the consistency of the one God whose will is the origin of the order of the created universe.

(b) *Trinitarian theology.* The counterpart in Christian theology of the physicists' GUT is the doctrine of the Trinity.[20] The Christians of the first centuries, as they reflected on the Church's experiences and understanding, came to recognise that they had known God in three fundamental ways. There

20. For more on the Trinity, see J. C. Polkinghorne, *Science and the Trinity*, SPCK/Yale University Press, 2004, especially ch. 4; *Exploring Reality*, ch. 5.

was the heavenly Father, Creator of the universe and the One who had given the Law to Moses in the clouds and thick darkness of Mount Sinai. God above us, one might say. There was the incarnate Son, Jesus Christ, sharing in and redeeming humanity, and making God's will and nature known in the plainest and most accessible terms through his human life in Palestine. God alongside us, one might say. There was the Holy Spirit, that divine presence at work in the human heart, bestowing gifts that matched individual personality and need (1 Corinthians 12:4–31). God within us, one might say. God known in three ways. Yet those early Christians also knew that they must hold on to the conviction that they had inherited from Judaism, that God is one. There is one divine will and purpose at work in the world. The divine drama has its three acts of creation, redemption and sanctification, attributed in much Christian speaking to the Father, Son and Spirit respectively, but there is one sole Author of the play. By the fourth century, struggling with the tensions between these different insights had led to the trinitarian doctrine that the one true God is constituted by the exchange of love between three divine Persons.

It is important to recognise that belief in the Holy Trinity was motivated by Christian experience in this way, and that it did not arise simply from rash and ungrounded metaphysical speculation about matters of deep mystery. What was predominantly involved was engagement with what the theologians call 'the economic Trinity', an evidenced-based argument from below. The adjective derives from the Greek word *oikonomia*, whose root meaning concerns the order of a household, in this case the household of the divinely created world. The experience that the Fathers relied on did not only come from the great revelatory events of Creation, Incarnation and Pentecostal empowerment, but it also arose from the ordinary worshipping life of the Church, which prayed to the Father through the Son and in the power of the Spirit, and whose characteristic acclamation of praise was, and remains, 'Glory to the Father, and to the Son, and to the Holy Spirit'.

This approach from below to the mystery of the divine being is based on an affirmation of theological realism, the conviction that encounters with the sacred reality of God will be trustworthy and not misleading. Such a belief was epigrammatically expressed by the twentieth-century Roman Catholic theologian, Karl Rahner, when he said that the economic Trinity (God as experienced) is the immanent Trinity (God in the mystery of the divine essence). As in Christology, so in trinitarian theology, the approach from below inevitably leads on to attempts to complete the discussion by conceptual analysis from above.

Two constraints have contained all orthodox conceptual thinking about the Trinity within certain limits. One constraint is a refusal to entertain the heresy called 'modalism'. This asserted that Father, Son and Spirit are simply labels for different perspectives on God, means of referring to different modes of encounter with a single divine Being. The Church found that modalism gave insufficient recognition to the distinctions between the Persons that were an actual feature of revelatory encounter. A key case was provided by the account of the baptism of Jesus (Mark 1:9–11 and parallels), in which the heavenly voice of the Father confirms the calling of the Son, upon whom the Holy Spirit descends in the form of a dove. The distinction of roles involved in this highly significant event implied that the Persons were not just different ways of looking at an undifferentiated God. True relationality requires true differentiation; the Father cannot be Father without the Son, nor the Son be Son without the Father. Yet the opposite extreme of tritheism, the picture of three separate gods, was equally unacceptable. Too loose and unintegrated an account—too 'social' a model of the household of God—would deny the divine unity that remains an essential Christian belief also. Obviously, the task of finding a middle way between these two rejected extremes is not an easy one.

Attempts to grapple with the problem led to the development of highly specialised and subtle theological language, particularly among the Greek Fathers, who felt that their linguistic resources were superior in this respect to what they considered to be the rather crude options available to the Latins. It is important to recognise that neither community would have conceived of a person in terms of the modern idea of an autonomous individual, so that a formula like 'Three Persons in One God' did not necessarily carry the suggestion of tritheism that it might do for a reader today. The three divine Persons are held to interpenetrate each other in the mutual exchange of love (a theological idea called 'perichoresis'), a concept that has no analogue in the case of three distinct human beings. Something of the character of the delicate distinctions to which the Fathers had recourse, can be seen by noting that they spoke of 'three *hypostaseis* in one *ousia*', despite the fact that until trinitarian refinements began to be formulated in the fourth century, *hypostasis* and *ousia* had been regarded as synonyms in Greek. One can get some feeling for the fine-grained character of this discourse by considering a possible English translation of the key phrase, 'three subsistences in one substance'. At times, the Fathers may have seemed to place an excessive degree of confidence in the success of a subtle elaboration of delicate detail, as when the differentiation between the Son and the Spirit was held to depend upon an elusive distinction between begetting and procession as expressions of their relationships to the Father. Perhaps the nearest analogy in physics to this kind of reliance on subtle and elaborated conceptual construction would be the baroque and only partially articulated eleven-dimensional edifice of M-theory, a conjectured foundation for string theory. All in all, however, one must confess once again that theology has proved less successful than science in attaining fully articulated theoretical understanding, a verdict that is scarcely surprising in view of the mysterious and infinite Reality about which theologians are striving to speak.

Trinitarian theologians have to hold on to their affirmation of the triune character of the one true God, even if they cannot fully articulate an explanation of how this can be the case. Once again, the threat of paradox cannot be dealt with simply by neglecting or denying the motivating experience. In holding on to a trinitarian understanding of God, theologians can take heart from the fact that their belief receives some support from collateral considerations. An influential contemporary book on trinitarian thinking has as its title *Being as Communion*.[21] One might paraphrase this as 'Reality is relational', an insight that certainly accords with an increasing scientific recognition of the relational character of the physical universe. The old-fashioned atomism that pictured isolated particles rattling around in the otherwise empty container of space has long been replaced by Gen-

21. J. Zizioulas, *Being as Communion*, St Vladimir's Seminary Press, 1985.

eral Relativity's integrated account of space, time and matter, understood to be combined in a single package deal. Quantum theory brought to light a remarkable form of entanglement between subatomic particles that have once interacted with each other (the so-called EPR effect), which implies that they remain effectively a single system however far they may subsequently separate spatially—a counterintuitive togetherness-in-separation that has been abundantly confirmed experimentally as a property of nature.[22] The physical world looks more and more like a universe that would be the fitting creation of the trinitarian God, the One whose deepest reality is relational.[23]

Finally, the picture of the three Persons, eternally united in the mutual exchange of love (the unique divine process of perichoresis), gives a profound insight into the meaning of that foundational Christian conviction that 'God is love' (1 John 4:8). A substantial revival of active interest in trinitarian thinking has taken place in recent years, inspired by just such considerations as these. I believe that the true 'Theory of Everything' is not superstrings, as physicists are sometimes moved bombastically to proclaim, but it is actually trinitarian theology.

22. See, Polkinghorne, *Quantum Theory*, ch. 5.
23. Polkinghorne, *Science and the Trinity*, ch. 3.

CHAPTER FIVE

Cousins

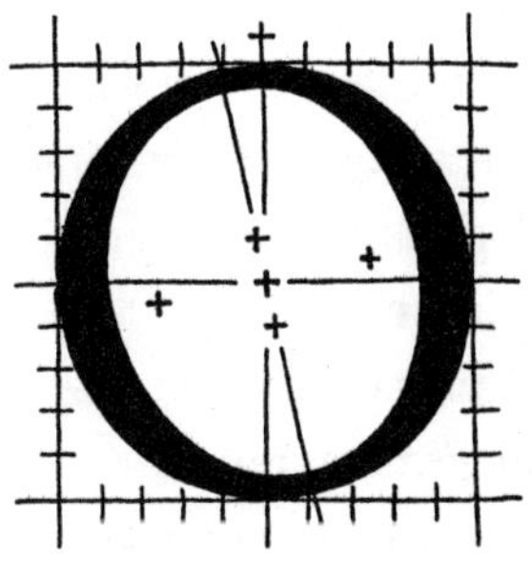

UR dissection of the truth-seeking strategies employed in science and theology has revealed significant underlying similarities between these two superficially different forms of rational enquiry, each concerned with its specific aspect of reality. In biology something rather similar happens when the comparative anatomists discover homologies between different forms of animal life. They then usually appeal to one of two possible explanations for the relationships they have discovered.

Orthodox Darwinian thinking will attribute the similarity to a common origin in the past, a primitive ancestor from which the two contemporary species have later diverged. Doubtless this is indeed frequently what has happened, but lately there has also been an increasing interest among theoretical biologists and palaeontologists in a second possible form of explanation. This suggests that when the organisation of biological entities complexifies, there are certain kinds of structures which preferentially emerge. One reason for this may lie in the character of the intrinsic self-organising properties of complex systems. This is an idea that has been developed particularly by Stuart Kauffman.[1] A second and related reason may be that the 'possibility world' explored by evolutionary processes may be more constrained than is often thought to be the case, because there may only be a limited number of basic structures that are both evolutionarily advantageous and readily biological accessible. The history of terrestrial life offers some encouragement to this idea. We know that eyes have developed several times independently in that history, in ways that certainly differ in some details, but which also manifest homologies in their basic structure (for example all use similar proteins, such as rhodopsin and the crystallins; cephalopods and mammals both have camera-eyes), and this would seem to support the idea of the constrained scope of biological possibility, forcing the independent repetition of successful strategies. Simon Conway Morris has written extensively on this convergent aspect of biological process, in which the same kinds of solutions to the problems of viable life forms frequently recur, even suggesting that if life exists elsewhere in the universe, its nature is likely to be more similar to its terrestrial counterpart than many might have supposed.[2] Beneath the apparent contingent diversity of life, there is believed to lie a deep substructure of universal principles shaping fruitful possibility.

1. S. Kauffman, *The Origins of Order*, Oxford University Press, 1993; *At Home in the Universe*, Oxford University Press, 1995.

2. S. Conway Morris, *The Crucible of Creation*, Oxford University Press, 1998; *Life's Solution*, Cambridge University Press, 2003.

Could one explain the cousinly relationships between the rational procedures of science and theology in ways analogous to these biological approaches? I suppose that appeal to a common ancestor would correspond to the thesis, quite often propounded,[3] that modern science owes religion a relationship of friendly gratitude since the latter historically provided the intellectual matrix that brought the former to birth. The argument supporting this claim runs as follows. The doctrine of creation of the kind that the Abrahamic faiths profess is such that it encourages the expectation that there will be a deep order in the world, expressive of the Mind and Purpose of that world's Creator. It also asserts that the character of this order has been freely chosen by God, since it was not determined beforehand by some kind of pre-existing blueprint (as, for example, Platonic thinking had supposed to be the case). As a consequence, the nature of cosmic order cannot be discovered just by taking thought, as if humans could themselves explore a noetic realm of rational constraint to which the Creator had had to submit, but the pattern of the world has to be discerned through the observations and experiments that are necessary in order to determine what form the divine choice has actually taken. What is needed, therefore, for successful science is the union of the mathematical expression of order with the empirical investigation of the actual properties of nature, a methodological synthesis of a kind that was pioneered with great skill and fruitfulness by Galileo. It was precisely this combination of theory and experiment that got modern science going in the seventeenth century, encouraged

3. R. Hooykaas, *Religion and the Rise of Modern Science*, Scottish Academic Press, 1973; S. Jaki, *The Road of Science and the Ways to God*, Scottish Academic Press, 1978; C. A. Russell, *Cross-Currents*, Inter-Varsity Press, 1985.

(it is suggested) by an ideology derived from viewing the universe as a free but orderly divine creation. Moreover, since the world is God's creation, it is a fitting duty for religious people to study it. In support of this thesis of the benign influence of religion on nascent science, one can note that it is indeed a fact that the pioneers of the scientific revolution were mostly persons of definite religious convictions, even if they may have had their problems with the authorities (Galileo) or with Christian orthodoxy (Newton). Those early scientists liked to say that God had written two books, the Book of Scripture and the Book of Nature. Both needed to be read, and when this was done aright there could be no contradiction between them, since the two had the same Author. It would have seemed very strange indeed to those pioneering figures to have suggested there was any real conflict between science and religion, enforcing a choice of which side to take. On this view, the 'primitive ancestor' of both modern science and modern theology was medieval *scientia*, understood in the general sense of Christian knowledge and insight.

A critical parting of the ways between these two forms of insight into reality began in the middle of the eighteenth century, when later generations of scientists became so flushed with the apparent success of mechanical argument that some began to make the triumphalist claim of the sufficiency of the scientific method on its own to yield all knowledge that was worth knowing, or even possible to know. Many scientists no longer interacted with the Church leadership in the way that Galileo had done, for they had come to treat theological insight as if it were of no relevance to the quest for true understanding. Yet by no means all scientists abandoned religious belief. For instance, Michael Faraday, James Clerk Maxwell and Lord Kelvin, three of the most outstanding figures in nineteenth-century physics, were all devout Christians.

The second kind of explanation offered for biological homologies appeals to the notion of deep underlying forms, whose universal patterns enable and shape the paths of fruitful development. The theological counterpart to this idea would be the doctrine of the *Logos*, the divine Word which is the fundamental source of the rational order of creation. John's gospel identifies the Word 'by whom all things were made' (1:3) with the incarnate Christ (1:14). In another passage of astonishing boldness, the Epistle to the Colossians identifies Christ as the one by whom 'all things in heaven and earth were created' and in whom 'all things hold together' (1:16–17). The *Logos* doctrine also speaks of the Word as enlightening everyone (John 1:9), an insight that can be appealed to for theological endorsement of the concept of critical realism.

These claims imply that the cousinly relationships that exist between different forms of creaturely truth-seeking endeavour derive ultimately from the fact that the universe was created as a true cosmos. It is an integrated world, whose deep intelligibility and consistency is a manifestation of the divine Word that lets be the whole of created reality (cf. Genesis 1: 'And God said 'Let there be . . .'). This in turn implies that religious people who are seeking to serve the God of truth should welcome all truth from whatever source it may come, without fear or reserve. Included in this open embrace must certainly be the truths of science. In the case of the scientists, the same insight implies that if they want to pursue the search for understanding through and through—a quest that it is most natural for them to embark upon—they will have to be prepared to go beyond the limits of science itself in the search for the widest and deepest context of intelligibility. I think that this further quest, if openly pursued, will take the enquirer in the direction of religious belief. It is a search for the *Logos.* In consequence, I believe that ultimately the cousinly relationships that we have investigated in this book find their most profound understanding in terms of that true Theory of Everything which is trinitarian theology.

Index